Spectroscopic Techniques for Dentistry Applications

Recent advances

Edited by
Ashutosh Kumar Shukla
Ewing Christian College, Prayagraj, India

IOP Publishing, Bristol, UK

ISBN 978-0-7503-3459-4 (ebook)
ISBN 978-0-7503-3457-0 (print)
ISBN 978-0-7503-3460-0 (myPrint)
ISBN 978-0-7503-3458-7 (mobi)

DOI 10.1088/978-0-7503-3459-4

Version: 20220301

IOP ebooks

British Library Cataloguing-in-Publication Data: A catalogue record for this book is available from the British Library.

Published by IOP Publishing, wholly owned by The Institute of Physics, London

IOP Publishing, Temple Circus, Temple Way, Bristol, BS1 6HG, UK

US Office: IOP Publishing, Inc., 190 North Independence Mall West, Suite 601, Philadelphia, PA 19106, USA

Spectroscopic Techniques for Dentistry Applications

Recent advances

IPEM–IOP Series in Physics and Engineering in Medicine and Biology

About the Series

The series in Physics and Engineering in Medicine and Biology will allow the Institute of Physics and Engineering in Medicine (IPEM) to enhance its mission to 'advance physics and engineering applied to medicine and biology for the public good'.

It is focused on key areas including, but not limited to:

- clinical engineering
- diagnostic radiology
- informatics and computing
- magnetic resonance imaging
- nuclear medicine
- physiological measurement
- radiation protection
- radiotherapy
- rehabilitation engineering
- ultrasound and non-ionising radiation.

A number of IPEM–IOP titles are being published as part of the EUTEMPE Network Series for Medical Physics Experts.

To my niece Aadvika.

Contents

Preface

This collection intends to highlight the applications of different spectroscopic techniques in dentistry. The six chapters in this volume have been written by experts. The team of contributing experts includes professional dentists in addition to academicians from different disciplines and thus the interdisciplinary character of this book has arisen in a natural way.

Chapter 1, by Zühre Akarslan, deals with laser-induced breakdown spectroscopy applications in dentistry. Saturnino Marco Lupi, Claudia Todaro, Benedetta Albini and Pietro Galinetto present Raman spectroscopy applications for dentistry in chapter 2. Chapter 3, by Monalisa Mishra, presents the applications of fluorescence spectroscopy in dentistry. Chapter 4, by Rajeev Ranjan, Diplina Barman, Prashant Gupta and Sandeep Kumar, reviews electron spin resonance spectroscopy applications for dentistry. Chapter 5, by Agata Niewczas and Przemysław M Płonka, presents emerging applications based on free radical generation by living tissues in normal conditions as well as in pathologies. Chapter 6, by Gunawan Indrayanto and Abdul Rohman, presents application of molecular spectroscopy and chemometrics in dentistry. This unique collection will be of interest to a wider audience, including dentists and biophysicists.

I sincerely thank T Michael Slaughter, Senior Commissioning Editor—Bioengineering, Medical Physics, Biophysics, and Emily Tapp, Commissioning Editor, of IOP Publishing for giving me an opportunity to present this book to readers. I wish to thank Robert Trevelyan and Phoebe Hooper, ebooks Editorial Assistants at IOP Publishing, for all the support they provided during the development of this project. Thanks are also due to the expert reviewers who have indirectly contributed to enhancing the quality of the content. I thank the expert authors for taking time out of their busy academic schedules to contribute to this volume.

Ashutosh Kumar Shukla
Prayagraj, India
August 2021

Editor biography

Ashutosh Kumar Shukla

Ashutosh Kumar Shukla has two decades of physics teaching and research experience. He has had numerous articles and review articles published in peer-reviewed journals. He has also authored textbooks and participated as an editor and contributor in many edited volumes published by reputed publishers, often prepared in collaboration with experts from different countries. He intends to continue pursuing his academic interests, which include ESR spectroscopy applications in different fields of societal importance.

List of contributors

Zühre Akarslan
Gazi University Faculty of Dentistry, Department of Oral and Dentomaxilofacial Radiology, Ankara, Turkey

Benedetta Albini
Department of Physics, University of Pavia, Pavia, Italy

Diplina Barman
Private Dental Practitioner, Chandannagar, West Bengal 712136, India

Pietro Galinetto
Department of Physics, University of Pavia, Pavia, Italy

Prashant Gupta
Department of Oral Medicine and Radiology, Dental Institute, RIMS Ranchi, Jharkhand 834009, India

Gunawan Indrayanto
Faculty of Pharmacy, Universitas Surabaya, Surabaya, East Java, Indonesia

Sandeep Kumar
Department of Public Health Dentistry, Dental Institute, RIMS Ranchi, Jharkhand 834009, India

Saturnino Marco Lupi
Department of Clinical Surgical, Diagnostic and Pediatric Sciences, University of Pavia, Pavia, Italy

Monalisa Mishra
Neural Developmental Biology Lab, Department of Life Science, NIT Rourkela, Rourkela, Odisha, India

Agata Niewczas
Chair and Department of Conservative Dentistry with Endodonthic, Medical University of Lublin ul. Chodźki 6, 20-093 Lublin, Poland

Przemysław M Płonka
Department of Biophysics, Faculty of Biochemistry, Biophysics and Biotechnology, Jagiellonian University in Kraków, ul. Gronostajowa 7, 30-387 Kraików, Poland

Rajeev Ranjan
Community Health Centre, Bero—Ranchi, Jharkhand 835202, India

Abdul Rohman
Center of Excellence, Institute for Halal Industry and System (IHIS), Gadjah Mada University, Yogyakarta 55281, Indonesia

Department of Pharmaceutical Chemistry, Faculty of Pharmacy, Gadjah Mada University, Yogyakarta 55281, Indonesia

Claudia Todaro
Department of Clinical Surgical, Diagnostic and Pediatric Sciences, University of Pavia, Pavia, Italy

IOP Publishing

Spectroscopic Techniques for Dentistry Applications
Recent advances
Ashutosh Kumar Shukla

Chapter 1

Laser-induced breakdown spectroscopy in dentistry

Zühre Akarslan

Laser-induced breakdown spectroscopy (LIBS) is an atomic emission spectroscopic technique. Elements and atomic species are quantified through spectrometric analysis of a laser-induced plasma. Teeth, jaw bones and oral soft tissues are suitable materials for LIBS analysis. LIBS has been used for the elemental analysis of teeth and dental amalgam, caries detection and caries removal, adhesive removal in orthodontic treatment, and oral and maxillofacial surgical applications. This chapter aims to give an overview of LIBS applications in dentistry in light of the current literature.

1.1 Introduction

Laser-induced breakdown spectroscopy (LIBS) is a technique that quantifies elements or atomic species through spectroscopic analysis of the radiation emitted from laser-induced plasmas.

In this technique, high-power laser pulses are delivered to the surface of the examined material in a short duration and a plasma spark is generated when the laser power passes over the threshold of the examined material. This produces a plasma which emits radiation while cooling. The emitted radiation from the plasma is collected using an optical fiber and delivered to a spectrometer. Then it is analyzed and the characteristics of the examined sample are given through the electronic temperature or electron density (figure 1.1) [1–3].

The main parts of LIBS include a laser system (commonly a neodymium:YAG (Nd:YAG) laser system), focusing and collection optics, a spectrometer and a data acquisition system (figure 1.2). It is easy to obtain the megawatt peak power levels required for plasma generation with an Nd:YAG laser (1064 nm). Spectrometers working at 200–1000 nm are preferred because the resolution of the spectrometer is important for achieving high sensitivity [2].

doi:10.1088/978-0-7503-3459-4ch1

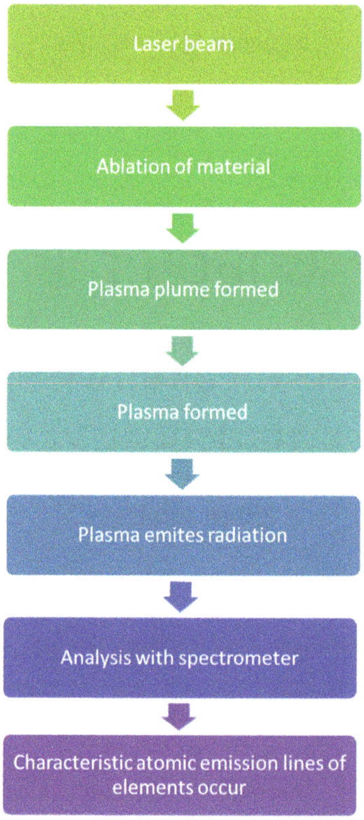

Figure 1.1. Basic workflow of LIBS.

LIBS is capable of performing multi-elemental analyses in solids, liquids and gases in a short time. It is an easy technique and provides real-time sample identification. No sample preparation or minimal sample preparation is necessary for the process and a large number of samples can be analyzed [1, 4]. However, it is difficult to obtain suitable matrix-matched standards. In addition, the precision and the detection limits of LIBS are not as good as the conventional techniques and high energy lasers can cause tissue damage. In addition, high energy lasers can cause tissue damage [2].

The first analytical use of laser-induced plasma was in the 1960s. With the development of Nd:YAG lasers, excimer lasers and other lasers with different wavelengths, LIBS applications have gained great attention in many disciplines [5, 6], such as archeology, physics, anthropology and biomedicine. Currently, the sensitivity of the technique can be increased by combining it with chemometrics and machine learning algorithms [7].

Teeth are a suitable material for LIBS analyses due to their structure. They are composed of enamel, dentine, pulp and cement. Enamel is the outer part of the tooth, covering the crown. It is exposed to the oral cavity and is in contact with oral fluids and other substances (figure 1.3). Dentine is the second layer of the tooth, under the enamel of the crown. The roots of the teeth are embedded in the jaws and thus they are not in contact with the oral cavity like the crown is in a healthy oral environment.

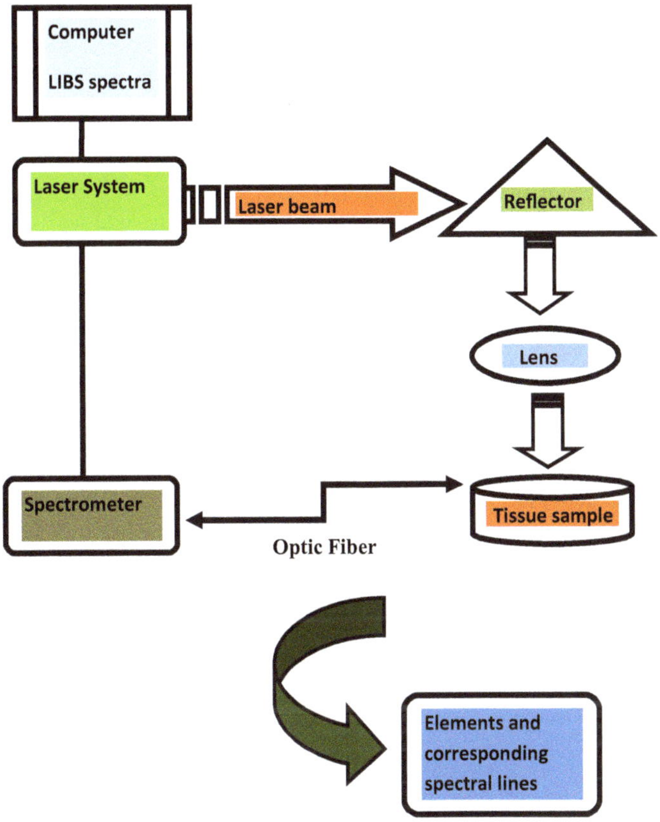

Figure 1.2. The main parts of LIBS.

The outer surface of the root is made of cement. Cement is in contact with the periodontal fibers which connect the teeth to the jaw bones. The inner side of the root also consists of dentine, similar to the crown. The innermost part of a tooth is the pulp tissue. This is located both in the crown and the root. The pulp tissue is mostly made up of blood vessels and nerves (figure 1.4).

The major component of enamel and dentine is hydroxyapatite ($Ca_{10}(PO_4)_6(OH)_2$). Enamel has a higher content of hydroxyapatite than dentine and is the hardest tissue of teeth. It contains 95% hydroxyapatite crystals, 4% water and 1% organic substance. Dentine contains 70% hydroxyapatite crystals, 10% water and 20% organic substance. The pulp tissue does not contain mineralized tissue.

Enamel is very hard due to the high hydroxyapatite concentration. This makes it a suitable material for laser applications. It has a lower ablation threshold and a higher reproducible laser-induced breakdown, compared to the other tissues. In addition, it has a high calcium (Ca) concentration. Ca has intense emission lines with low detection limits, therefore it can be detected easily using the LIBS technique [8].

Human teeth contain essential and toxic elements. The elements which do not have any role in the physiological structure of teeth are potentially toxic.

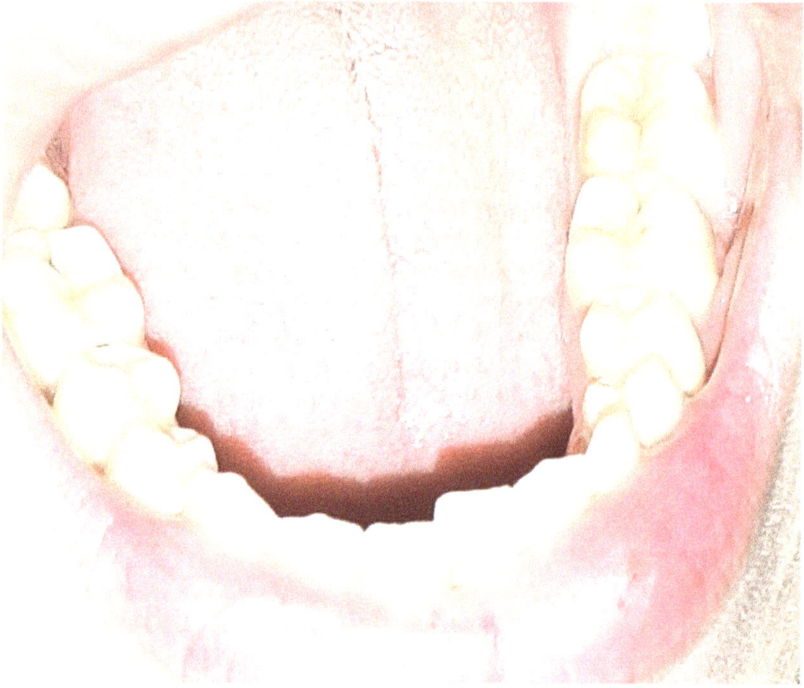

Figure 1.3. Crowns of permanent teeth in the oral cavity.

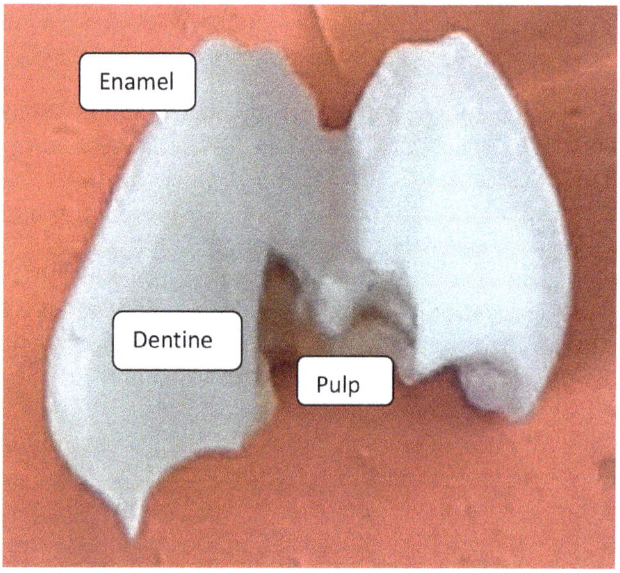

Figure 1.4. Enamel, dentine and pulp cavity.

Table 1.1. Elements of enamel, dentine and pulp.

Tissue	Element	Researcher, year, reference
Enamel	Cn, Ca, Na, K, P, H, C, O, Fe, Zn	Rohde *et al* (2017) [10]
	Ca, P, O, H, Mg, Li, Ba, Zn, P, Na	Samek *et al* (2001) [11]
	Ca, P, O, H, Zn, Sr, Fe, Al, B, Ba, Pb, Ni, Li, Ag, As, Se, Nb, Hg	Kang *et al* (2004) [12]
	Ca, Mg, Na, Cu, P, K, Cl, Mn, Cu, Zn, Sr, Pb, C	Derise *et al* (1974) [13, 14]
	Ca, P, Mg, Na, K, Sr, C, H, Zn, O	Gazmeh *et al* (2015) [15]
Dentine	Ca, Na, K, P, N, H, C, O, Fe, Zn	Rohde *et al* (2017) [10]
	Ca, P, O, H, Zn, Sr, Fe, Al, B, Ba, Pb, Ni, Li, Ag, As, Se, Nb, Hg	Kang *et al* (2004) [12]
	Ca, Mg, Na, Cu, P, K, Cl, Mn, Cu, Zn, Sr, Pb, Co	Derise *et al* (1974) [13, 14]
	Ca, P, O	Moshonov *et al* (2005) [16]
	Al, Pb, Sn, Li, As, Cd, Ca, P, Mg, Na, S, K, Sr, Zn, Ba, Fe, B, Ti, Mn, Cr, Ni, Cu, Co, Se	Fernández-Escudero *et al* (2020) [9]
Pulp	C, Mg, Fe, Ca, Na, H, N, K, O, Cn	Rohde *et al* (2017) [10]
	H, C, N, O, Na, S, Cl, H, C, N, O, Na, P, S, Cl, K	Woodard *et al* (1986) [17]
	Ca, Mg, Na, Cu, P, K, Cl, Mn, Cu, Zn, Sr, Pb, Co	Derise *et al* (1974) [13, 14]
	Ca, P, O	Moshonov *et al* (2005) [16]

Note: aluminum (Al), arsenic (As), boron (B), barium (Ba), cadmium (Cd), calcium (Ca), cobalt (Co), chrome (Cr), chlorine (Cl), copernicium (Cn), copper (Cu), hydrogen (H), iron (Fe), potassium (K), magnesium (Mg), manganese (Mn), sodium (Na), nickel (Ni), phosphorus (P), selenium (Se), silver (Ag), stannum (Sn), strontium (Sr), sulfur (S), titanium (Ti), lithium (Li), lead (Pb), oxygen (O), canadium (V), zinc (Zn).

Mineral exchanges occur in coronal dentine during the oral functioning of teeth. Thus the structure of the dentine changes with age. It has been shown that both the concentration of essential elements, including boron (B), barium (Ba), potassium (K), strontium (Sr), sulfur (S) and magnesium (Mg), and toxic elements, including lead (Pb), lithium (Li) and stannum (Sn), in permanent teeth increased with age. In addition, the concentration of Pb and K in the coronal dentin of premolars and molars showed high variations among young and elderly individuals [9]. Details of the elements present in the teeth are given in table 1.1.

1.2 Lasers and LIBS in dentistry

Lasers started to be used in dentistry shortly after their introduction in the medical field. Currently there are different laser types with different wavelengths for various oral applications. The laser type, tissue composition, and the amount of air, water, and blood in the examination area are factors considered in laser applications [18]. Details are given in table 1.2.

The first attempt at the use of lasers in dentistry was in the 1960s. A group of researchers tested a long-pulse ruby laser to remove caries from teeth, however, this method caused thermal damage and could not achieve success [19]. Later, lasers with short and ultra-short pulses and high repetition rates were introduced and the

Table 1.2. Lasers used in different dental procedures.

Device type	Wavelength	Dental application
CO_2 laser	10600 nm	Surgery
Erbium:YAG laser (Er:YAG laser) and chromium:YSGG laser (CR:YSGG laser)	2780–2940 nm	Surgery, endodontics and cavity preparation
Neodymium:YAG laser (Nd:YAG laser)	1064 nm	Surgery, endodontics and periodontology
Diode laser	810–980 nm	Surgery, endodontics and periodontology

efficacy of these lasers was tested in dental applications. These lasers reduced the thermal damage to teeth structures, provided a favorable operation time for the removal of the dental tissue and also provided feedback on the procedure through the radiation emitted by the laser-induced plasma [8]. Since then, LIBS has been used for several purposes in dentistry. These can be grouped as follows:

1. Analysis of elements in teeth.
2. Caries detection and removal.
3. Elemental analysis of dental amalgam.
4. Adhesive removal in orthodontics.
5. Surgical applications.

1.2.1 Analysis of elements in teeth

Human dentition is composed of primary and permanent teeth. The primary teeth start to erupt in the oral cavity when a baby is 6–8 months old. Generally, the maxillary and mandibular incisors erupt first. The first permanent teeth start to erupt at 6 years of age in the majority of children. The primary maxillary and mandibular incisors are the first teeth that are replaced by their permanents. The first permanent molars also usually erupt at 6 years of age. The other teeth are then replaced by their permanents and at ages 11–12 the second permanent molars begin to erupt [20]. The maxillary and mandibular third molars are the last teeth to erupt. They usually erupt at the age of 20 [21] (figure 1.5). Thus the primary and permanent teeth erupt at different ages and stay in the oral cavity for different periods. The permanent teeth are not replaced by any other dentition. The number and duration of primary teeth in the oral cavity are lower compared to the permanent teeth. Thus they are exposed for a shorter period and to a smaller amount of chemicals and toxic materials.

The tooth crown is exposed to the oral cavity. Thus it is in contact with the oral environment. Eating, drinking, chewing, tooth-brushing and saliva affect the enamel. Enamel is made up of organic and inorganic materials and a small amount of water. The majority of the enamel is inorganic, including hydroxyapatite crystals, minerals and a small number of various metals. The organic part of the enamel is composed of protein and fat, etc [22].

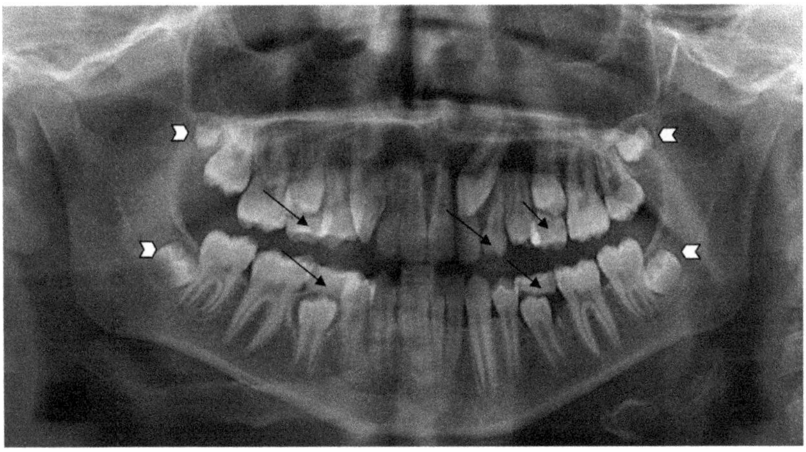

Figure 1.5. Panoramic radiograph of a child with primary and permanent teeth. All the primary second molars and a single primary canine tooth are present in the oral cavity (black arrows). The other primary teeth have been replaced by their permanents. The permanent first and second molars are erupted in the oral cavity, while the third molars are in the development stage in the jaws (white arrowhead).

Khalid *et al* evaluated the variations in concentrations of Ca, Fe, Sr, Zn and Pb elements present in the enamel, dentine and cement of primary teeth using a constant laser fluence of 2.6 J cm^{-2}. They used an Nd:YAG laser (1064 nm, 10 ns) for the evaluation of plasma parameters and elemental analysis. They assessed the electron temperature and electron number density from the tooth plasma at various fluences. Both parameters showed an increasing trend up to a certain value of laser fluence, i.e. 2.6 J cm^{-2}. When the laser fluence value was increased to 3.9 J cm^{-2}, a shielding effect occurred and thus a decreasing trend was observed. When the laser fluence was increased up to a maximum value of 10.5 J cm^{-2}, a saturation phenomenon governed by a self-regulating regime occurred, therefore the plasma parameters showed insignificant changes. Based on these findings, they reported that laser fluence controlled the plasma parameters. It was found that enamel had the highest amount of micro-minerals, following by dentine and cement [23].

Using the LIBS technique, Samek *et al* investigated the possible effect of environmental factors on the elemental structure of primary and permanent teeth. They found that Al was present in the primary and the permanent teeth, both with and without dental restorations. Thus they concluded that Al from toothpaste can penetrate teeth [24].

The irrigation of root canals is important for successful endodontic treatment of a tooth. Using the LIBS method, Küçükkaya Eren *et al* investigated the mineral content of the root canal dentine of permanent teeth after applying different irrigation solutions *ex vivo*. These researchers irrigated the root canals of teeth with 5.25% sodium hypochlorite (NaOCl) for 1 h, 17% ethylenediaminetetra acetic acid (EDTA) for 2 min, and 5.25% NaOCl for 1 h + 17% EDTA for 2 min. It was found that the CA/P ratio decreased, in contrast to the increase in Na levels in the canals which were irrigated with the NaOCl and NaOCl+EDTA groups [25].

1.2.2 Caries detection and removal

Caries is a common disease of human teeth (figure 1.6). Both children and adults are affected by this disease. Although the prevalence of caries varies among countries, the risk factors and the developmental stages are the same [26].

As, discussed above, teeth are composed of enamel, dentine, pulp and cement, and the major component of enamel and dentine is hydroxylapatite ($Ca_{10}(PO_4)_6(OH)_2$]) [27]. Caries develops when molecular changes start in the hydroxylapatite crystals in the enamel. The apatite crystals dissolve and the tooth loses Ca, P and other ions. This process is called demineralization. The pH of the oral cavity is an important factor for caries development. Caries occur when teeth are exposed to acids. Acids dissolve enamel according to the following reaction:

$$Ca_{10}(PO_4)_6(OH)_2\ (s) + 8H^+\ (aq) \rightarrow 10Ca^{2+}\ (aq) + 6HPO^2{}_{4-}\ (aq) + 2H_2O\ (l).$$

Caries detection and removal are important as an untreated carious lesion can lead to pain, tooth loss, and even jaw infection. The visual–tactile method, radiography, transillumination, quantitative light-induced fluorescence, DIAGNOdent, fiber-optic transillumination and electrical methods are used for the detection of caries. However, none of these methods has 100% sensitivity and specificity [28].

Studies have been conducted on LIBS and caries detection. An early study was carried out by a group of researchers who assessed LIBS in the differentiation of the

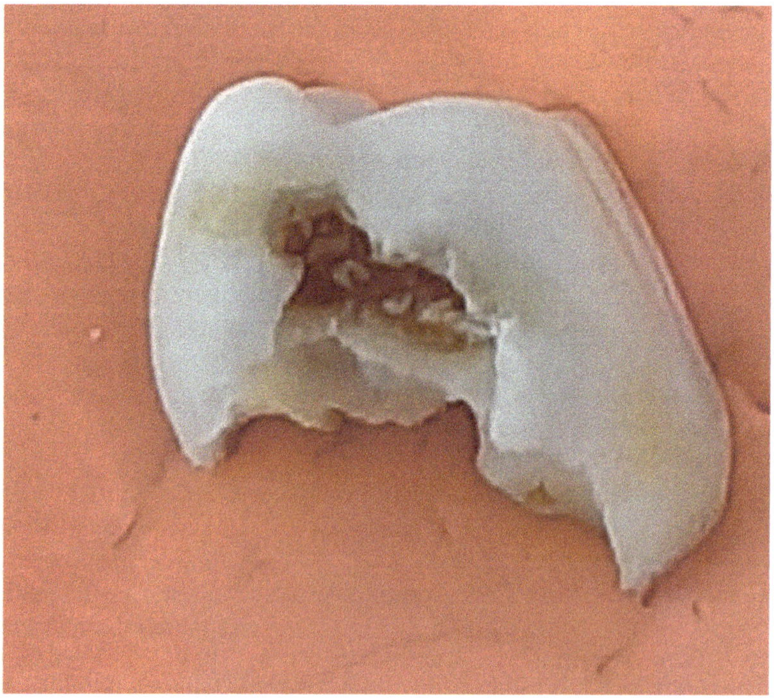

Figure 1.6. A carious lesion in an extracted primary tooth section.

healthy and carious parts of teeth, and in cavity preparation. They found that LIBS provided a good distinction between the spectra of healthy and carious teeth. The decrease in the Ca and Na elements due to demineralization in the carious part of teeth was seen with this technique. They also prepared a dental cavity with a 40 ps Nd:YLF laser by the ablation of the enamel and dentin [29].

Samek *et al* reported that the LIBS technique was a simple and applicable technique for the identification of carious and sound tooth sections. They used the Mahalanobis distance pattern recognition algorithm in discriminant analysis to evaluate the LIBS spectra. They found that using LIBS with discriminant analysis was useful for caries detection during the laser drilling process (Er:YAG laser) with short laser pulses under *in vivo* and *in vitro* circumstances [11].

Thareja *et al* assessed the presence of Na, Sr and Ca elements in healthy and carious parts of teeth using LIBS in conjunction with discriminate analysis. This research reported that Ca was reduced in the carious part of the tooth and LIBS was capable of nearly 100% identification of healthy and carious teeth [30].

Sing and Rai examined the elements in healthy and carious tooth sites using the prominent transitions of the atomic lines from LIBS. They reported that Ca, Mg, Cu, Zn, Sr, Ti, C, P, H, O, Na and K elements were present in teeth. Compared to the healthy sites of a tooth, the Ca and P levels decreased in carious lesions. In contrast, the carious site had a higher content of Mg, Cu, Zn, Sr, C, Na and K. Also, they found that healthy tooth sites had lower H and O levels [27].

Sasazawa *et al* designed an LIBS system for real-time *in vivo* analysis of carious and healthy enamel. They used a hollow optical fiber for the transmission of Q-switched Nd:YAG laser light for LIBS and infrared Er:YAG laser light for tooth ablation. They found that focusing on the emission peaks of Zn in the ultraviolet region in addition to expanding the spectral region under analysis to ultraviolet light increased the sensitivity of the system for enamel caries detection [22].

A recent study evaluated the efficiency of LIBS in the differentiation of healthy and carious tooth structures. The results of this study indicate that healthy and carious enamel contained similar elements, but the number of some of the elements decreased in the carious part. They found a large amount of Ca and P and small amounts of Mg, Na, K, Sr, C, H, Zn and O in the healthy and carious parts of the teeth. However, the amounts of these elements changed in the carious parts of the teeth [15].

Gazmeh *et al* conducted another study on the discrimination of the healthy and carious parts of teeth using LIBS and partial least squares discriminant analysis for the emission lines of P, Ca, Mg, Zn, K, Sr, C, Na, H and O elements. They concluded that this method was successful in the prediction of the state of the tooth [31].

Khalid *et al* assessed the variations in elements in primary teeth using LIBS. Primary teeth have the same elements as permanent teeth but are exposed to the oral cavity for a shorter time compared to permanent teeth. The results of their study indicated that primary teeth with caries had enhanced concentrations of Pb, Sr, Zn and Fe [23].

1.2.3 Elemental analysis of dental amalgam

Amalgam is a restorative material used in dentistry. It is resistant, easy to prepare and easy to manipulate in the oral cavity [3]. It is a metallic alloy and has two main components, liquid mercury and a matrix, each of which makes up 50% of the amalgam. The liquid mercury consists of Hg and the matrix consists of Ag–Cu–Sn. Sometimes the matrix may also contain Zn. First, the liquid mercury is mixed with the matrix and a malleable paste is formed. Then this paste is placed into the prepared dental cavity where it hardens and gains its final structure. The main disadvantage of amalgam restorations is the presence of the element Hg in the amalgam. Hg vapor is released from dental amalgam restorations and Hg^{2+} penetrates the matrix of the teeth. Hg is a heavy metal which is toxic for humans and accumulates in tissues [32]. An amalgam restoration in an extracted primary tooth is shown in figure 1.7.

Teeth slowly release the elements they absorb. Therefore toxic element analysis can be performed with LIBS under *ex vivo* circumstances [5]. Researchers have investigated the penetration of amalgam elements into dental tissues and the calibration of analytical methods.

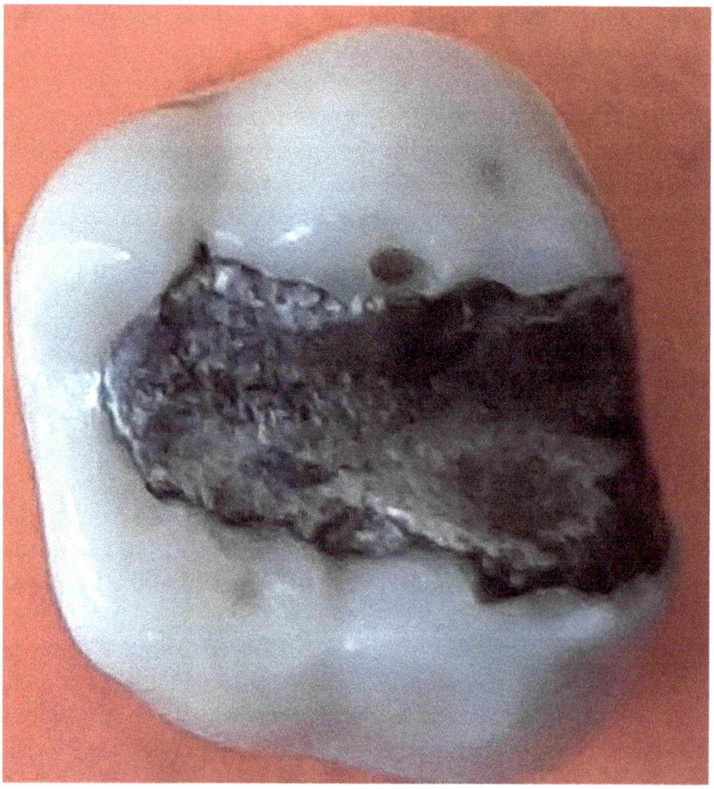

Figure 1.7. An amalgam restoration in an extracted primary tooth.

Castellon *et al* assessed the accuracy of calibration curves at various delay times for optimal calibration during the analysis of Hg in amalgam. Using LIBS they performed the elemental analysis of dental amalgams containing different mercury concentrations in an Ag–Cu–Sn matrix. They found that the electron number density and temperature increased when the Hg concentration in the amalgam alloy increased. In addition, they suggested an optimal delay of 3000 ns [3].

A recent study assessed the penetration of amalgam elements in teeth using LIBS. The results indicate that Hg, Ag, Cu and Sn penetrated into the dental matrix of teeth with amalgam restorations and the penetration occurred mostly near the restoration site. In addition, amalgam diffusion decreased from the restoration site to the inner parts of the teeth [15].

Femtosecond (ultra-short) lasers are also used for the ablation process in LIBS. This type of laser reduces matrix effects. Tooth samples can be collected using the femtosecond-LIBS technique. A recent study evaluated the penetration of Hg and the other amalgam elements in teeth using femtosecond-LIBS. The use of this technique promoted a less selective ablation and decreased the dependence on the material matrix, which helped in the determination of the elements in amalgam restorations and teeth. The researchers performed the experiment with an amplified Ti:sapphire laser system. They used both primary and permanent teeth with amalgam restorations. They reported that qualitative assessments of the elements could be made with a spatial resolution of less than 100 mm. The spectra of the amalgam restorations and teeth were different. The emission lines of Ag, Hg, Cu and Sn predominated in the amalgam spectrum, and Ca and Mg predominated in the tooth spectrum. The amalgam penetration depth was similar in the primary and permanent dentition. It was determined that Hg diffused more deeply in the permanent teeth compared to the primary teeth. This may be due to the longer duration of amalgam restorations in permanent dentition, as permanent teeth remain in the oral cavity for a long time compared to primary teeth [32].

Extracted teeth contain copper, zinc and mercury. It was found that the coronal part of a tooth crown with an amalgam restoration contained the highest mercury concentrations compared to the other sections. Copper concentrations were higher, but zinc concentrations were lower in teeth without restorations compared to restored teeth. Thus amalgam restorations may be a source of zinc [33].

1.2.4 Adhesive removal in orthodontics

Orthodontics treats malocclusions and incorrect spacing between teeth. It provides normal alignment of teeth and helps to maintain oral health. Brackets are the main materials used during orthodontic treatment. In the beginning, the brackets have to be bonded to the crowns of the teeth with resins. For this, the enamel is etched with acids, rinsed and dried. Then the adhesive is applied to the bracket pad and the brackets are positioned and polymerized with a visible light curing device. The brackets stay on the teeth until the end of the orthodontic treatment, when they are removed using the procedure of bracket debonding. During this procedure the adhesive resin is removed from the tooth surface. Debonding should be performed very carefully to avoid iatrogenic damage to the tooth surface. Laser ablation has been shown to be an alternative method to the conventional techniques for debonding [34, 35]. The combination of laser ablation with analytical techniques helps the clinician to identify the material and thus control the procedure [36].

Gomez *et al* assessed the removal of adhesive on teeth after bracket debonding by using laser radiation at 355 nm (the third harmonic wavelength of a Q-switched Nd: YAG laser) monitored with LIBS. The LIBS spectra showed distinct differentiation between the elements of the enamel and the adhesive. The characteristic lines of Ca were dominant in the emission spectrum of the enamel while Si and Ti were dominant in the adhesive with a good signal-to-noise ratio. They found that this method was helpful for the assessment of proper removal of the adhesive from enamel [37].

1.2.5 Surgical applications

With the advent of small and portable lasers, laser surgery has gained great attention due to its advantages compared to conventional techniques in oral and maxillofacial surgery. Lasers are used for soft and hard tissue cutting, treatment of nerve injuries, periodontal regenerative procedures [38] and the treatment of patients with coagulation deficiency [39].

Laser beam penetration differs between different types of lasers. The penetration of the laser beam is related to the power level, pulse repetition rate, pulse width and energy density [40]. Ultra-short lasers have suitable parameters to be used in surgical procedures, however, nanosecond-lasers are more preferred during surgery as they reduce the thermal stress and provide more precise ablation in LIBS studies [8].

Laser surgery reduces the stress level of the patient, lowers post-operative complication rates and provides better healing of the operation area with minimal invasive applications compared to conventional techniques. Also, the ablation provides the coagulation of the blood vessels cut during surgery. This allows the surgeon to perform the surgery on a dry and clear site [10, 41]. However, laser surgery has an increased risk of iatrogenic damage to anatomical structures because the operator cannot know the actual penetration depth of the laser beam exactly. Thus they cannot control the ablated tissue type effectively. LIBS can overcome this disadvantage and make it possible to select the tissue which is going to be ablated. A recent study investigated LIBS efficacy in differing oral soft and hard tissues. Cortical and cancellous bone, nerve, mucosa, enamel, dentine and pulp were examined. Further, the tissues were classified using emission-ratio-based analysis and linear discriminant analysis. It was found that calcium:carbon-based ratios had 100% sensitivity and 99.91% specificity for the differentiation of hard and soft tissues. Thus LIBS was capable of accurate differentiation of the hard and soft tissues. The linear discriminant analysis was successful in the differentiation of the investigated samples, except for the enamel–dentine combination [10].

Early research in this area has shown that LIBS is very successful in the differentiation of cartilage and bone tissue, but is not good at discriminating adipose–fat tissue, compared to the cartilage and bone tissue pairing [42, 43].

In the above studies a spectrometer with a low resolution was used. Recently, a study was conducted with a spectrometer with a higher resolution and more successful results were achieved [44]. The identification and preservation of nerves in the operation site are important for successful surgery. Thus it would be very helpful if a system could distinguish such tissues during laser surgery. Studies have evaluated the efficacy of LIBS on this issue.

The main elements of the nerves and adipose tissues are carbon, oxygen, hydrogen and nitrogen. It was reported that carbon, oxygen and potassium significantly affected the differentiation of the adipose and nerve tissues. Although the elemental analysis of adipose tissue and nerves with LIBS suggested that the elemental compositions of the two tissues are similar, they had different element concentrations. The concentration of carbon in fat tissue from the pure carbon emission lines was higher than that in the nerve tissue. The nerve tissue spectra presented higher Na, K and O values than the adipose tissue. Thus measurement of the concentration of these elements with LIBS is helpful for the identification of adipose and nerve tissue during oral surgery [44].

1.3 Conclusion

After the development of lasers the LIBS technique became available for many applications. LIBS can be used for elemental analysis of teeth and dental amalgam, caries detection and caries removal, adhesive removal in orthodontic treatment, and surgical applications in dentistry. With the continuing advances in the field, this system has the potential for further applications.

References

[1] Anabitarte F, Cobo A and Lopez-Higuera J M 2012 Laser-induced breakdown spectroscopy fundamentals applications and challenges *ISRN Spectrosc.* **2012** 285240

[2] Singh V K and Rai A K 2011 Prospects for laser-induced breakdown spectroscopy for biomedical applications a review *Lasers Med. Sci.* **26** 673–87

[3] Castellon E, Pacheco Martinez P, Alvarez J, Bredice F, Borges F, Muniz M V, Sanchez-Ake C, Palleschi V and Sarmiento R 2018 Elemental analysis of dental amalgams by laser-induced breakdown spectroscopy technique *Spectroc. Acta* B **149** 229–35

[4] Alvira F C, Ramirez Rozzi F and Bilmes G M 2010 Laser-induced breakdown spectroscopy microanalysis of trace elements in *Homo sapiens* teeth *Appl. Spectrosc.* **64** 313–19

[5] Singh V K, Kumar V and Sharma J 2015 Importance of laser-induced breakdown spectroscopy for hard tissues (bone teeth) and other calcified tissue materials *Lasers Med. Sci.* **30** 1763–78

[6] Tognoni E, Palleschi V, Corsi M and Cristoforetti G 2002 Quantitative micro-analysis by laser-induced breakdown spectroscopy—a review of the experimental approaches *Spectroc. Acta* B **57** 1115–30

[7] Wang Q, Xiangli W, Teng G, Cui X and Wei K 2021 A brief review of laser-induced breakdown spectroscopy for human and animal soft tissues pathological diagnosis and physiological detection *Appl. Spectrosc. Rev.* **563** 221–41

[8] Gaudiuso R, Melikechia N, Abdel-Salamc Z A, Palleschid M A, Motto-Rose V and Busser B 2019 Laser-induced breakdown spectroscopy for human and animal health: a review *Spectroc. Acta* B **152** 123–48

[9] Fernández-Escudero A C, Legaz I, Prieto-Bonete G, López-Nicolás M, Maurandi-López A and Pérez-Cárceles M D 2020 Aging and trace elements in human coronal tooth dentine *Sci. Rep.* **10** 9964

[10] Rohde M, Mehari F, Klämpfl F, Adler W, Neukam F W, Schmidt M and Stelzle F 2017 The differentiation of oral soft- and hard tissues using laser induced breakdown spectroscopy—a prospect for tissue specific laser surgery *J. Biophotonics* **10** 1250–61

[11] Samek O, Telle H H and Beddows D 2001 Laser-induced breakdown spectroscopy a tool for real-time *in vitro* and *in vivo* identification of carious teeth *BMC Oral Health* **1** 1

[12] Kang D, Amarasiriwardena D and Goodman A H 2004 Application of laser ablation-inductively coupled plasma-mass spectrometry (LA-ICP-MS) to investigate trace metal spatial distributions in human tooth enamel and dentine growth layers and pulp *Anal. Bioanal. Chem.* **378** 1608–15

[13] Derise N L, Ritchey S J and Furr A K 1974 Mineral composition of normal human enamel and dentin and the relation of composition to dental caries. I: Macrominerals and comparison of methods of analyses *J. Dent. Res.* **53** 847–52

[14] Derise N L and Ritchey S J 1974 Mineral composition of normal human enamel and dentin and the relation of composition to dental caries. II: Microminerals *J. Dent. Res.* **53** 853–8

[15] Gazmeh M, Bahreini M, Tavassoli S H and Asnaashari M 2015 Qualitative analysis of teeth and evaluation of amalgam elements penetration into dental matrix using laser induced breakdown spectroscopy *J. Lasers Med. Sci.* **6** 67–73

[16] Moshonov J, Stabholz A, Bar-Hilel R and Peretz B 2005 Chemical analysis and surface morphology of enamel and dentin following 96μ CO_2 laser irradiation versus high speed drilling *J. Dent.* **33** 427–32

[17] Woodard H Q and White D R 1986 The composition of body tissues *Br. J. Radiol.* **59** 1209–18

[18] Saydjari Y, Kuypers T and Gutknecht N 2016 Laser application in dentistry irradiation effects of Nd:YAG 1064 nm and diode 810 nm and 980 nm in infected root canals—a literature overview *BioMed Res. Int.* **2016** 8421656

[19] Goldman L, Hornby P, Meyer R and Goldman B 1964 Impact of the laser on dental caries *Nature* **203** 417

[20] Parner E T, Heidmann J M, Vaeth M and Poulsen S 2001 A longitudinal study of time trends in the eruption of permanent teeth in Danish children *Arch. Oral Biol.* **46** 425–31

[21] Shaweesh A I 2016 Timing of clinical eruption of third molars in a Jordanian population *Arch. Oral Biol.* **72** 157–63

[22] Sasazawa S, Kakino S and Matsuura Y 2015 Optical-fiber-based laser-induced breakdown spectroscopy for detection of early caries *J. Biomed. Opt.* **20** 065002

[23] Khalid A, Bashir S, Akram M and Hayat A 2015 Laser-induced breakdown spectroscopy analysis of human deciduous teeth samples *Lasers Med. Sci.* **30** 2233–8

[24] Samek B O, Telle H H, Morris G W, Liska M and Kaiser J 1999 Quantitative analysis of trace metal accumulation in teeth using laser-induced breakdown spectroscopy *Appl. Phys.* A **69** 179–82

[25] Küçükkaya Eren S, Uzunoğlu E, Sezer B, Yılmaz Z and Boyacı İ H 2018 Mineral content analysis of root canal dentin using laser-induced breakdown spectroscopy *Restor. Dent. Endod.* **43** e11

[26] Guido J A, Martinez Mier E A, Soto A, Eggertsson H, Sanders B J, Jones J E, Weddell J A, Villanueva Cruz I and Anton de la Concha J L 2011 Caries prevalence and its association with brushing habits water availability and the intake of sugared beverages *Int. J. Paediatr. Dent.* **21** 432–40
[27] Singh V K and Rai A K 2011 Potential of laser-induced breakdown spectroscopy for the rapid identification of carious teeth *Lasers Med. Sci.* **26** 307–15
[28] Gomez J 2015 Detection and diagnosis of the early caries lesion *BMC Oral Health* **15** S3
[29] Willms L, Herschel A, Niemz M H and Pioch T 1996 Preparation of dental hard tissue with picosecond laser pulses *Lasers Med. Sci.* **11** 45–51
[30] Thareja R K, Sharma A K and Shobha Shukla S 2008 Spectroscopic investigations of carious tooth decay *Med. Eng. Phys.* **30** 1143–8
[31] Gazmeh M, Bahreini M and Tavassoli S H 2015 Discrimination of healthy and carious teeth using laser-induced breakdown spectroscopy and partial least square discriminant analysis *Appl. Opt.* **154** 123–31
[32] Bello L T, Ana P A, Santos Jr D, Krug F J, Zezell D M, Vieira Jr N D and Samad R E 2017 Mercury amalgam diffusion in human teeth probed using femtosecond LIBS *Appl. Spectrosc.* **71** 659–69
[33] Brigato R C, Costa L M, da Costa M R, Assis N M and Kubo C H 2009 Mercury copper and zinc concentrations in extracted human teeth *Arch. Environ. Occup. Health* **64** 266–9
[34] Lu Y F, Takai M, Komuro S, Shiokawa T and Aoyagi Y 1994 Surface cleaning of metals by pulsed-laser irradiation in air *Appl. Phys.* A **59** 281–8
[35] Tsunemi A, Hagiwara K, Saito N, Nagasaka K, Miyamoto Y, Soto O and Tashiro H 1996 Complete removal of pain from metal surface by ablation with a TEA CO_2 laser *Appl. Phys.* A **63** 435–9
[36] Castillejo M *et al* 2002 Analytical study of the chemical and physical changes induced by KrF laser cleaning of tempera paints *Anal. Chem.* **74** 4662–71
[37] Gómez C, Palma J C and Costela A 2017 On-line laser radiation controlled to the removal of adhesive on teeth after bracket debonding *Laser Ther.* **261** 25–30
[38] Kirpalani T and Dym H 2020 Role of piezo surgery and lasers in the oral surgery office *Dent. Clin. North Am.* **64** 351–63
[39] Strauss R A and Fallon S D 2004 Lasers in contemporary oral and maxillofacial surgery *Dent. Clin. North Am.* **48** 861–88
[40] Cobb C M 2006 Lasers in periodontics a review of the literature *J. Periodontol.* **77** 545–64
[41] Gáspár L 1994 The use of high-power lasers in oral surgery *JCLMS* **12** 281–5
[42] Kanawade R, Mehari F, Knipfer C, Rohde M, Tangermann-Gerk K, Schmidt M and Stelzle F 2013 Qualitative tissue differentiation by analysing the intensity ratios of atomic emission lines using laser induced breakdown spectroscopy (LIBS) prospects for a feedback mechanism for surgical laser systems *Spectrochim. Acta* B **87** 175–81
[43] Kanawade R, Mahari F, Klämpfl F, Rohde M, Knipfer C, Tangermann-Gerk K, Adler W, Schmidt M and Stelzle F 2015 Qualitative tissue differentiation by analysing the intensity ratios of atomic emission lines using laser induced breakdown spectroscopy (LIBS) prospects for a feedback mechanism for surgical laser systems *J. Biophotonics* **8** 153–61
[44] Mehari F, Rohde M, Kanawade R, Knipfer C, Adler W, Klämpf F, Stelzle F and Schmidt M 2016 Investigation of the differentiation of *ex vivo* nerve and fat tissues using laser-induced breakdown spectroscopy (LIBS): prospects for tissue-specific laser surgery *J. Biophotonics* **9** 1021–32

Chapter 2

Raman spectroscopy for dentistry applications

Saturnino Marco Lupi, Claudia Todaro, Benedetta Albini and Pietro Galinetto

2.1 Introduction

Raman spectroscopy (RS) is a non-destructive optical method based on the analysis of the inelastic scattering of light diffused by a certain molecular aggregate. The Indian scientist Sir Chandrasekhara Venkata Raman discovered the effect later named after him during his studies on the scattering of the light from liquids and the results were published in 1928 (Raman and Krishnan 1928). Thanks to this discovery Raman won the Nobel Prize in 1930. RS is able to quantitatively and qualitatively characterize a large variety of materials, including solids, liquids and gases. In particular, micro-RS allows the inspection of solid samples, even nano-structured samples, at the surface level and/or at a depth of a few tens of microns depending on their optical properties (Ferraro 2005, Nascimento 2018).

Raman observed a diffused radiation field from benzene irradiated by monochromatic visible light. In addition to the elastic scattering, the so-called Rayleigh component, he revealed a feeble signal with a lower frequency with respect to the incident frequency due to an inelastic scattering process. Furthermore, Raman understood that the frequency difference between the incident and inelastic scattered radiation was dependent only on the irradiated material (Raman and Krishnan 1928) and this was the result of an energy transfer from light to a vibrational (or rotational) energy reservoir of the material. The plot of the inelastic scattered light versus the wavenumbers gives rise to the Raman spectrum. The Raman signal is characteristic for each type of molecule, depending uniquely on the chemical composition and the bond strength and symmetry. When the atoms are arranged in a crystalline manner the symmetry constraints become stronger and the Raman effect can be described, introducing the phonon concept. From a quantum-mechanical point of view, referring to the simplified scheme reported in figure 2.1, the Raman effect can be described as follows: a photon of energy equal to $h\nu$ is impinging on the molecule and the photon energy is lower than the energy

doi:10.1088/978-0-7503-3459-4ch2

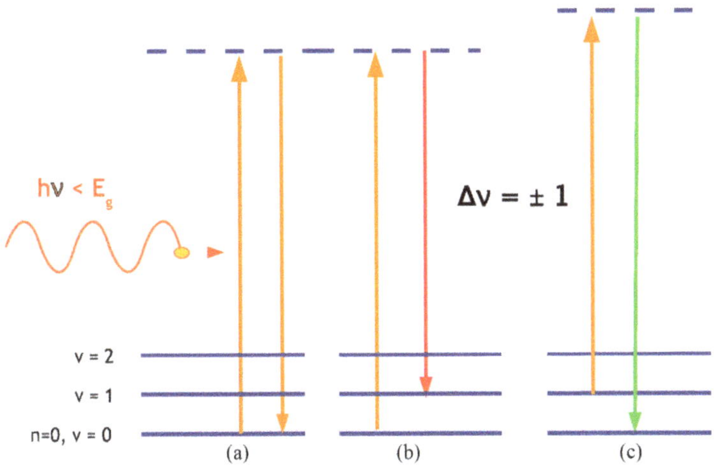

Figure 2.1. The Raman effect from a quantum-mechanical point of view.

gap between the fundamental and excited electronic energy levels of the molecule. One can imagine that the electron is excited to a so-called virtual state, characterized by a very short lifetime, from where it can decay, returning to the same electronic and vibrational state (figure 2.1(a)). This give rise to Rayleigh scattering, with diffused light with the same energy as the irradiating energy. A very small fraction of light (10^{-6} or more than the incident light) can experience a different path: the electron as in Rayleigh scattering is excited to a virtual state but the decay process ends on a different vibrational level (figure 2.1(b)) with an energy loss of the emitted photon equal to the vibrational energy gap between adjacent levels. This gives rise to the so-called Stokes Raman scattering. Nevertheless, another process is possible (figure 2.1(c)): the incident photon excites an electron residing on an excited vibrational level and then when it decays from the virtual state to the electronic ground level, in addition to the Rayleigh scattering and the Stokes Raman scattering, it can decay on a lower vibrational level with the emitted photon having higher energy with respect the incident one. This is the so-called anti-Stokes Raman scattering, a much less likely event at ambient temperature because the electrons are largely located on the vibrational ground level.

The Raman spectrum represents a kind of fingerprint of the investigated material (Ko *et al* 2006, Krafft and Sergo 2006). In the spread of the use of RS, technological advances have played a crucial role, first the advent of lasers in the 1960s and then in the 1970s the coupling of microscopy to Raman spectrometers and the use of CCD detectors. Today RS is used in an extremely wide range of research and analytical fields, from materials science to biomedicine, from pharmaceuticals to electronics, from forensic science to pipeline monitoring and more (Nafie 2020, Dresselhaus *et al* 2010, Galinetto *et al* 2018, Kudelski 2008, Gao *et al* 2021, Vandenabeele and Edwards 2018). This is due to the interplay of different advantages: it is a non-invasive and non-destructive method that generally avoids sample degradation or alteration; it is extremely specific and quite sensitive; it is simple in execution as it does not require complex equipment or procedures; and portable instruments are now available and the signal can be easily driven by optical fibers allowing remote measurements (Talari *et al* 2017, Movasaghi *et al* 2008).

The implementation of microscopes in Raman spectrometers allows one to realize Raman sampling with a high lateral resolution down to the diffraction limit, around 0.5 μm. Also, a good resolution in depth (some microns) is obtained when measuring transparent materials and realizing quasi-confocal sampling in this way. Concerning the sensitivity, it is of importance to mention surface enhanced RS (SERS). In SERS a huge increase of the normal Raman signal is produced from a molecular aggregate when it is typically close to a colloidal preparation of noble metal nanoparticles (Schlücker 2014). The molecules close to the nanoparticles are stimulated by an enhanced local field thanks to the plasmonic resonances with laser light. This leads to the possibility to reveal the signal even from a single molecule. SERS has further boosted the use of RS, allowing it to also work with visible light on biological samples due to the simultaneous presence of enhanced signal and quenching of luminescence, usually a detrimental interference in RS, particularly in biological materials. In some cases, normal RS is performed on biological samples using infrared light. In this case a Fourier transformed spectrometer is used leading to so-called FT-RS (Rohman *et al* 2020).

This chapter presents the applications of RS in the dental field through a complete review of relevant publications.

2.2 Demineralization and remineralization of hard dental tissues; caries diagnosis; demineralization due to whitening

RS allows us to evaluate the variation of the organic and inorganic structure of the enamel and dentin of teeth after exposure to thermal and mechanical stimuli, demineralizing and remineralizing agents, adhesive resins and whiteners (Manuel Toledano, *et al* 2014e, Toledano-Osorio *et al* 2018, Sender *et al* 2020).

2.2.1 RS and whitening techniques

RS was used to evaluate the effects of the following whitening agents on the molecular structure of enamel and its degree of reflectivity (Liang *et al* 2020). Among the various agents were hydrogen peroxide 35% (Cesar *et al* 2009), carbamide peroxide 10% (Redha *et al* 2019), and fluoride and calcium in addition to 10% carbamide peroxide (Cavalli *et al* 2011). RS was used to evaluate the effects of whitening agents such as sodium perborate activated with a Nd:YAG laser and diodes on the molecular structure of dentin by means of an intracoronal approach (Sağlam *et al* 2015). These studies found that there were no noticeable changes in the structural integrity of the enamel exposed to hydrogen peroxide, but the exposure even to low concentrations of carbamide peroxide caused the deterioration of the dentinal collagen matrix.

2.2.2 RS and remineralization

Studies show that cluster analysis through RS can evaluate the level and quality of remineralization of dentin subjected first to different etching techniques and then to different bonding techniques. Thermal and mechanical cycles were investigated in order to simulate normal chewing function and some types of parafunction (Toledano-Osorio *et al* 2018, Manuel Toledano *et al* 2014a, 2014c, 2015). In these *in vitro* studies the goal was to find the more appropriate type and frequency of vibration to stimulate remineralization of the bonded dentin. Evidence showed that dentin's functional remineralization was obtained after the application of stimuli.

Fourier transformed RS associated with laser treatment was used to obtain fluorescent-free Raman signals from biological materials, allowing the molecular structure of hydroxyapatite crystals formed by the incorporation of the molecules after deposition and APF treatments to be identified (Tsuda and Arends 1997).

Many studies have analyzed these characteristics using different molecules, including zinc chloride, zinc oxide, saliva, chlorhexidine, phosphate-based solution, bioactive glass, polyaspartic acid and sodium trimetaphosphate with resin-based zinc cement, vegetable extracts based on anacardaceae, chitosan and polymethylmethacrylate based silver-filled nanoparticles (Osorio *et al* 2014, Osorio *et al* 2018, Zhang *et al* 2018, Moreira *et al* 2017, Osorio *et al* 2016, Sauro *et al* 2011, Iafisco *et al* 2018, Toledano *et al* 2014d). All the studies listed above concluded that the analyzed materials were, at different levels and conditions, capable of inducing dentin remineralization and improvement in the strength of the collagen matrix.

2.2.3 RS and protocols of prevention of demineralization

Through RS it has been possible to evaluate the effects of pretreatments with coating materials on dentin subjected to erosion. Specifically, after the application of a layer based on carbon nanotubes/graphene oxide or nanohydroxyapatite, or a combination of both, the teeth were subjected to different abrasive techniques and analyzed using RS in order to identify the quantity and quality of the material residue on the surface (Soares *et al* 2020). Through RS it is possible to evaluate the inhibitory effects on demineralization of the dentin by means of surface irradiation with a CO_2 laser (Vieira *et al* 2015).

Fourier transformed RS was used to study the changes to the enamel of deciduous teeth following cycles of radiotherapy and cycles of pH variation. The spectroscopy bands related to the organic components (carbonate and phosphate) were evaluated (de Sá Ferreira *et al* 2016). One group of deciduous molars was irradiated, mimicking a radiotherapy cycle of 5 weeks and 2 days, the other group was not irradiated. The spectroscopy bands related to the organic components of carbonate and phosphate were evaluated and the results showed a decrease in the organic component of the enamel. This study suggests that the clinician should adopt preventive strategies to reduce the degradation of the enamel surface.

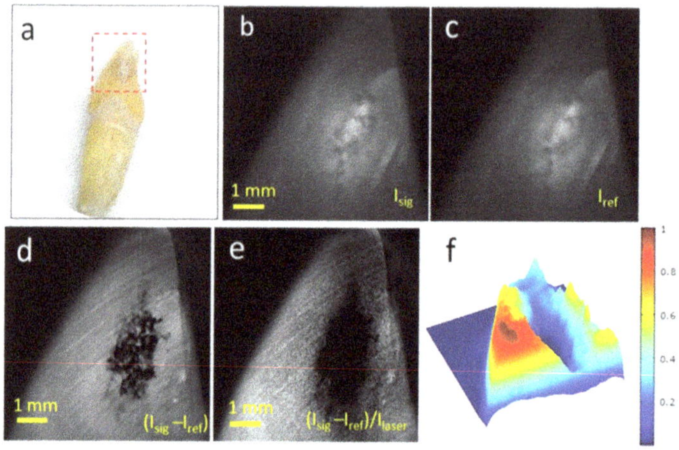

Figure 2.2. (a) A photograph of an incisor with a white lesion taken using a cell phone camera. Full-scale images covering the lesion when the band pass filter was tuned to (b) the Raman signal peak and (c) the fluorescence reference. Fluorescence subtracted images (d) before and (e) after laser intensity correction. The laser intensity corrected Raman image is replotted in (f) in 3D format for additional viewpoints. Reproduced with permission from Yakubu *et al* 2018, copyright The Optical Society.

2.2.4 RS and caries diagnosis

Early diagnosis of the hard dental tissue demineralization process remains a challenging task in dentistry. The characteristics of spectroscopy are particularly interesting in the diagnosis of carious lesions, already from the earliest stages of development (white spot), since RS and the Fourier transform allow the loss of intensity of hydroxyapatite to be identified (Tetschke *et al* 2018). In particular, RS allows not only clinically evident carious lesions to be diagnosed, but also the carious lesions that are not recognizable by traditional clinical examination thanks to the high sensitivity and specificity of the assay of the mineralized tissues of the tooth (Yakubu *et al* 2018). This study demonstrated that the full-scale Raman images based on the mineral intensity distribution can be used to separate the enamel from dentin, as well as normal enamel from decayed enamel. More importantly, the Raman images may identify dental caries that are not visually detectable and it is a promising technique for caries detection for clinicians.

2.3 Evaluation of dental adhesives and polymerization of resins

2.3.1 RS and dentin/composite interface evaluation

Micro-RS can identify the characteristic spectra of collagen, hydroxyapatite and adhesive (Yakubu *et al* 2018), making it a fundamental tool for understanding the chemical and physical changes at the level of the hybrid dentin/adhesive layer. Spectroscopy can be used to assess the degree of polymerization of various adhesives at the adhesive–demineralized dentin/enamel interface. Furthermore, it is possible to establish the degree of penetration into the tubules and the thickness of the adhesive layer. All this makes it possible to effectively compare the various adhesive techniques (Santini and Miletic 2008, Al-Nabulsi *et al* 2019, Fu *et al* 2020, Siqueira *et al* 2019, Cardenas *et al* 2019, Costa *et al* 2019, Huilcapi *et al* 2020, Okulus *et al* 2017, Toledano *et al* 2014b, Daood *et al* 2018, Loguercio *et al* 2015, Oliveira *et al* 2017, Pongprueksa *et al* 2015, Almahdy *et al* 2015, Talungchit *et al* 2014).

The topography and chemical composition of the dentinal surface treated with non-thermal atmospheric plasma (NTAP) was characterized by atomic force microscopy and by Raman confocal spectroscopy in order to evaluate the modifications in the enzymatic activity and the consequences on the microtensile strength and the thickness of the adhesive layer (Ayres *et al* 2018). Other studies evaluated the chemical and physical changes undergone by the hybrid layer following exposure to various substances. In particular, the effects following exposure to an acid and aqueous solution with 2% chlorhexidine were evaluated (Ayres *et al* 2018). RS was used to evaluate the effectiveness of the different bonding systems on tetracycline-stained teeth (Ayres *et al* 2018).

2.3.2 RS and the chemical and physical properties of resins and cements

Micro-RS is used widely to evaluate the speed and the degree of polymerization (the conversion degree or rate of conversion) of various resin-based cements and the consequent chemical and physical properties. Numerous studies have made it possible to evaluate the following:

- Glass ionomer polymers synthesized in CO_2 under supercritical conditions (with sufficient temperature and pressure CO_2 behaves as a gas and solid simultaneously) and methanol as the co-solvent (Moshaverinia *et al* 2008).
- Hardening reaction through transesterification of silicate resin and its potential as a biocompatible endodontic cement (Portella *et al* 2014).
- Timing and degree of polymerization of resins containing type 1 photocatalysts (Randolph *et al* 2014).
- Physiochemical characterization of fiber reinforced composites (FRCs) (Bonon *et al* 2016).
- The effects of temperature and heating time on epoxy resins and calcium silicate based cements (Atmeh and AlShwaimi 2017).
- The effects of the dissolution capacity of different primer systems on the degree of polymerization and on the bond strength between composite cements and FRCs (Khan *et al* 2018).

2.4 RS and the characterization of ceramic materials for dental prostheses

Micro-RS and confocal RS can be used to evaluate the molecular structure and physical and chemical properties of zirconia, allowing an effective comparison between the various manufacturing procedures (Nakonieczny *et al* 2016, Sorrentino *et al* 2019). In addition, RS is used to evaluate the adhesion between the zirconia or metal substrate and the feldspar ceramic overlay (Durand *et al* 2012, Yao *et al* 2013). The bond strength between the adhesive layer and the zirconia or lithium disilicate can also be measured using RS (Gutierrez *et al* 2020, Cardenas *et al* 2017). RS represents a non-invasive method to verify the effects of mechanical and chemical stresses on the molecular structure of zirconia parts (Sorrentino *et al* 2019, Ramos *et al* 2015, Spies *et al* 2020).

2.5 RS characterization of dental implants

Surface-RS represents a non-invasive analysis method for the evaluation of implants made of zirconia or titanium (Khmaladze *et al* 2015, Lupi *et al* 2020a, 2020b). Biocompatibility parameters of the various implant surface treatments can be evaluated. In particular, the speed of clot formation, the apposition of bone cells and the degree of ossification were studied in relation to the following surface treatments: nanocrystalline zinc oxide on titanium implants, nanocrystalline silicates on zirconia implants, nanocrystalline titanium dioxide on titanium implants, nano-crystalline graphene oxide on titanium implants, and nanoparticles of calcium phosphate and biodegradable nanoparticles based on copper and silver (Sousa *et al* 2014, Penkov *et al* 2016, Burtscher *et al* 2019, Hung *et al* 2016, Ren *et al* 2017, Pelaez-Vargas *et al* 2011). The mechanical strength of the zirconia and titanium implants was assessed (Zhang *et al* 2020, Bozzini *et al* 2008, Bozzini *et al* 2011).

The biocompatibility and health of peri-implant tissues is influenced by the presence of contaminating substances on the implant surface and by the degree of corrosion of the oxidized layer that forms on the surface itself (Lutin *et al* 2015, Bozzini *et al* 2011, Bozzini *et al* 2008).

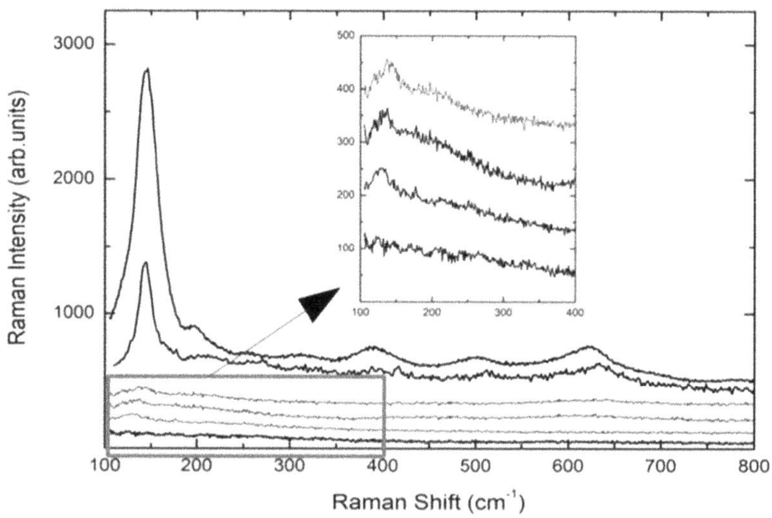

Figure 2.3. Raman spectra obtained from Ti–6Al–4V alloy and commercially pure implants. From the bottom: machined alloy and commercially pure implants; blasted and double etched alloy implants; blasted, double etched and anodized alloy implants; blasted, double etched, anodized and cold plasma treated alloy implants; blasted and double etched commercially pure implants; and blasted, double etched and anodized commercially pure implants. In the inset the low energy region for the Ti–6Al–4V alloy samples is reported in order to give evidence of the Raman signal at around 145 cm^{-1}. Reproduced from Lupi *et al* (2020b).

2.6 RS and evaluation of grafting materials

RS has been demonstrated to be able to evaluate the organic and inorganic structure of the regenerated bone. Deproteinized bovine bone was investigated using RS (Charwat-Pessler *et al* 2015). Biocompatible and biodegradable optical fiber polymers (PDLLA) were investigated with confocal RS to evaluate the osteoinductive potential comparing different production methods and to evaluate cell adhesion (Charwat-Pessler *et al* 2015). The osteoinductive potential and cytotoxicity of materials that combine the biocompatibility characteristics of bio-ceramics with the resistance of carbon nanotubes were investigated (Rodrigues *et al* 2016). Also nanogel grafts based on hydroxyapatite carbonate were evaluated (Adachi *et al* 2020). RS was used to confirm the antibacterial and osteoinductive ability of scaffolds containing pure phosphorite (Krishnamurithy *et al* 2019).

The RS has also been used to evaluate the protective barrier properties of some types of membrane used for bone regenerative procedures. The osteoinductive potential of a boron-modified bioactive nanoparticle glass containing bilayered membrane was demonstrated using Fourier transform RS (Rad *et al* 2019). Polyurethane membranes whose surfaces were treated with different waterproofing procedures have been investigated (Sheikh *et al* 2016). Also, bone healing after the application of a norbixin membrane scaffold was evaluated with and without laser photo-biomodulation (Alves *et al* 2018).

RS was used to evaluate the biocompatibility and determine the speed of apposition, and the quality and the thickness of the apatite layer that forms in contact with the bio-ceramic cements used for sealing endodontic canals based on calcium silicate and calcium hydroxide (Ranjkesh *et al* 2016, Gandolfi *et al* 2009, 2010, Toledano *et al* 2017, Uslu *et al* 2021). Moreover, RS was used to evaluate whether the addition of radiopaque materials (bismuth) leads to alterations in the *in vitro* biological behavior of ceramic materials (Mohn *et al* 2010).

2.7 Evaluation of organic materials: bone, periodontal ligament, periodontal intrasulcular fluid

2.7.1 RS and crevicular fluid

RS can be used to study crevicular fluid (Camerlingo et al 2014). In particular, some studies investigated the biochemical variations of the composition of the fluid itself and of the collagen of the periodontal ligament in the case of dental elements subjected to orthodontic forces (Jung *et al* 2014, d'Apuzzo *et al* 2017, Perillo *et al* 2020).

2.7.2 RS and tumors of the mouth

RS can be used for diagnosing cancer and targeting cancer cells by spectroscopic analysis of saliva and oral secretions, and by the identification of tumor cells through the use of gold and silver nanoparticles (Zhao *et al* 2015, Singh *et al* 2016, Ou *et al* 2014).

2.8 RS and orthodontics

RS has been used to evaluate the mechanical properties and changes due to the permanence in the oral cavity of two different materials for orthodontic aligners (Condo' *et al* 2018). The RS was applied for the evaluation of the adhesion of orthodontic brackets in fluorotic teeth (Huilcapi *et al* 2020). RS allows the evaluation of the mechanical and tribological properties of graphene oxide/silver nanoparticle coated medical grade nickel–titanium (NiTi) alloy for orthodontic use, concluding that nanocomposite coated NiTi alloy demonstrated an improved mechanical strength and a reduced friction coefficient, i.e. more favorable features for biomedical applications (Rokaya *et al* 2019).

2.9 RS and various applications

RS can be used in the bioengineering field for the realization of substrates for the release of drugs, including antibiotic molecules (Schofield *et al* 2006, Padmanabhan *et al* 2018). In the case of a gene mutation, RS can help identify the quality and quantity of macromolecules included in the hydroxyapatite crystals; for example, in the case of mutation in the gene that codes for MMP-20 (Prajapati *et al* 2016).

Raman stable isotope probing (SIP), a combination of RS and D_2O labeling, is used to evaluate carbon assimilation and metabolic activity at the cellular level. This technique is particularly useful in the analysis of the metabolism of bacterial communities, for example, of the oral cavity (Wang *et al* 2016). Together with scanning electron microscopy (SEM), RS was used to evaluate which is the best substrate for the adhesion of diamonds on cutters in order to guarantee better performance and less wear (Jackson *et al* 2007).

References

Adachi T *et al* 2020 *In vivo* regeneration of large bone defects by cross-linked porous hydrogel: a pilot study in mice combining micro tomography, histological analyses, Raman spectroscopy and synchrotron infrared imaging *Materials* **13** 4275

Almahdy A, Koller G, Festy F, Bartsch J W, Watson T F and Banerjee A 2015 An MMP-inhibitor modified adhesive primer enhances bond durability to carious dentin *Dental Mater.* **31** 594–602

Al-Nabulsi M, Daud A, Yiu C, Omar H, Sauro S, Fawzy A and Daood U 2019 Co-blend application mode of bulk fill composite resin *Materials* **12** 2504

Alves A M M, de Miranda Fortaleza L M, Maia Filho A L M, Leão Ferreira D C, da Costa C L S, Freitas Viana V G, Verde Santos J Z L, de Oliveira R A, de Meira Gusmão G O and Silva Soares L E 2018 Evaluation of bone repair after application of a norbixin membrane scaffold with and without laser photobiomodulation (λ 780 nm) *Lasers Med. Sci.* **33** 1493–504

Atmeh A R and AlShwaimi E 2017 The effect of heating time and temperature on epoxy resin and calcium silicate-based endodontic sealers *J. Endod.* **43** 2112–8

Ayres A P, Bonvent J J, Mogilevych B, Soares L E S, Martin A A, Ambrosano G M, Nascimento F D, Van Meerbeek B and Giannini M 2018 Effect of non-thermal atmospheric plasma on the dentin-surface topography and composition and on the bond strength of a universal adhesive *Eur. J. Oral Sci.* **126** 53–65

Bonon A J, Weck M, Bonfante E A and Coelho P G 2016 Physicochemical characterization of three fiber-reinforced epoxide-based composites for dental applications *Mater. Sci. Eng. C: Mater. Biol. Appl.* **69** 905–13

Bozzini B, Carlino P, D'Urzo L, Pepe V, Mele C and Venturo F 2008 An electrochemical impedance investigation of the behaviour of anodically oxidised titanium in human plasma and cognate fluids, relevant to dental applications *J. Mater. Sci. – Mater. Med.* **19** 3443–53

Bozzini B, Carlino P and Mele C 2011 Electrochemical behaviour and surface characterisation of Zr exposed to an SBF solution containing glycine, in view of dental implant applications *J. Mater. Sci. – Mater. Med.* **22** 193–200

Burtscher S, Krieg P, Killinger A, Al-Ahmad A, Seidenstücker M, Latorre S H and Bernstein A 2019 Thin degradable coatings for optimization of osteointegration associated with simultaneous infection prophylaxis *Materials* **12** 3495

Camerlingo C, d'Apuzzo F, Grassia V, Perillo L and Lepore M 2014 Micro-Raman spectroscopy for monitoring changes in periodontal ligaments and gingival crevicular fluid *Sensors* **14** 22552–63

Cardenas A F M, Armas-Veja A, Rodriguez Villarreal J P, de Siqueira F S F, Muniz L P, Campos V S, Reis A and Loguercio A D 2019 Influence of the mode of application of universal adhesive systems on adhesive properties to fluorotic enamel *Braz. Oral Res.* **33** e120

Cardenas A M, Siqueira F, Hass V, Malaquias P, Gutierrez M F, Reis A, Perdigão J and Loguercio A 2017 Effect of MDP-containing silane and adhesive used alone or in combination on the long-term bond strength and chemical interaction with lithium disilicate ceramics *J. Adhes. Dent.* **19** 203–12

Cavalli V, Azevedo Rodrigues L K, Paes-Leme A F, Silva Soares L E, Martin A A, Berger S B and Giannini M 2011 Effects of the addition of fluoride and calcium to low-concentrated carbamide peroxide agents on the enamel surface and subsurface *Photomed. Laser Surg.* **29** 319–25

Cesar I C R, Silva Soares L E, Alves L P, Martin A A, Munin E and Suzy Liporoni P C 2009 Fourier transform-Raman and reflectance studies on dental enamel bleached with hydrogen peroxide activated using a light-emitting diode-laser system *Photomed. Laser Surg.* **27** 913–9

Charwat-Pessler J, Musso M, Petutschnigg A, Entacher K, Plank B, Wernersson E, Tangl S and Schuller-Götzburg P 2015 A bone sample containing a bone graft substitute analyzed by correlating density information obtained by x-ray micro tomography with compositional information obtained by Raman microscopy *Materials* **8** 3831–53

Condo' R *et al* 2018 Mechanical properties of Two generations of teeth aligners: change analysis during oral permanence *Dental Mater. J.* **37** 835–42

Costa A C, Sabóia V, Marçal F, Sena N, De Paula D, Ribeiro T and Feitosa V 2019 *In vitro* evaluation of experimental self-adhesive orthodontic composites used to bond ceramic brackets *Materials* **12** 419

d'Apuzzo F, Perillo L, Delfino I, Portaccio M, Lepore M and Camerlingo C 2017 Monitoring early phases of orthodontic treatment by means of Raman spectroscopies *J. Biomed. Opt.* **22** 1–10

Daood U, Tsoi J K H, Neelakantan P, Matinlinna J P, Omar H A K, Al-Nabulsi M and Fawzy A S 2018 *In vitro* assessment of ribose modified two-step etch-and-rinse dentine adhesive *Dental Mater.* **34** 1175–87

Dresselhaus M S, Jorio A and Saito R 2010 Characterizing graphene, graphite, and carbon nanotubes by Raman spectroscopy *Annu. Rev. Condens. Matter Phys.* **1** 89–108

Durand J-C, Jacquot B, Salehi H, Margerit J and Cuisinier F J G 2012 Confocal Raman microscopy and SEM/EDS investigations of the interface between the zirconia core and veneering ceramic: The influence of a liner and regeneration firing *J. Mater. Sci. – Mater. Med.* **23** 1343–53

Ferraro J 2005 *Introductory Raman Spectroscopy* 2nd edn (Amsterdam: Elsevier)

Fu C, Deng S, Koneski I, Awad M M, Akram Z, Matinlinna J, Pichika M R, Daood U and Fawzy A S 2020 Multiscale *in-vitro* analysis of photo-activated riboflavin incorporated in an experimental universal adhesive *J. Mech. Behav. Biomed. Mater.* **112** 104082

Galinetto P, Albini B, Bini M and Mozzati M C 2018 Raman spectroscopy in zinc ferrites nanoparticles *Raman Spectroscopy* (Rijeka: InTech)

Gandolfi M G, Ciapetti G, Perut F, Taddei P, Modena E, Rossi P L and Prati C 2009 Biomimetic calcium-silicate cements aged in simulated body solutions. Osteoblast response and analyses of apatite coating *J. Appl. Biomater. Biomech.* **7** 160–70

Gandolfi M G, Ciapetti G, Taddei P, Perut F, Tinti A, Cardoso M V, Van Meerbeek B and Prati C 2010 Apatite formation on bioactive calcium-silicate cements for dentistry affects surface topography and human marrow stromal cells proliferation *Dental Mater.* **26** 974–92

Gao M, Liu S, Chen J, Gordon K C, Tian F and McGoverin C M 2021 Potential of Raman spectroscopy in facilitating pharmaceutical formulations development—an AI perspective *Int. J. Pharm.* **597** 120334

Gutierrez M F, Perdigão J, Malaquias P, Cardenas A M, Siqueira F, Hass V, Reis A and Loguercio A D 2020 Effect of methacryloyloxydecyl dihydrogen phosphate-containing silane and adhesive used alone or in combination on the bond strength and chemical interaction with zirconia ceramics under thermal aging *Oper. Dent.* **45** 516–27

Huilcapi M, Armas-Vega A, Cardenas A F M, Araujo L C R, Ocampo J B, Bandeca M C, de Siqueira F S F and Loguercio A 2020 Effect of surface treatments on the adhesive properties of metallic brackets on fluorotic enamel *Dental Press. J. Orthod.* **25** 59–67

Hung W-C, Chang F-M, Yang T-S, Ou K-L, Lin C-T and Peng P-W 2016 Oxygen-implanted induced formation of oxide layer enhances blood compatibility on titanium for biomedical applications *Mater. Sci. Eng. C: Mater. Biol. Appl.* **68** 523–9

Iafisco M, Esposti L D, Ramírez-Rodríguez G B, Carella F, Gómez-Morales J, Ionescu A C, Brambilla E, Tampieri A and Delgado-López J M 2018 Fluoride-doped amorphous calcium phosphate nanoparticles as a promising biomimetic material for dental remineralization *Sci. Rep.* **8** 17016

Jackson M J, Sein H, Ahmed W and Woodwards R 2007 Novel diamond-coated tools for dental drilling applications *J. Med. Eng. Technol.* **31** 81–93

Jung G B, Kim K-A, Han I, Park Y-G and Park H-K 2014 Biochemical characterization of human gingival crevicular fluid during orthodontic tooth movement using Raman spectroscopy *Biomed. Opt. Express* **5** 3508–20

Khan A A, Al-Kheraif A A, Mohamed B A, Perea-Lowery L, Säilynoja E and Vallittu P K 2018 Influence of primers on the properties of the adhesive interface between resin composite luting cement and fiber-reinforced composite *J. Mech. Behav. Biomed. Mater.* **88** 281–7

Khmaladze A, Kuo S, Kim R Y, Matthews R V, Marcelo C L, Feinberg S E and Morris M D 2015 Human oral mucosa tissue-engineered constructs monitored by Raman fiber-optic probe *Tissue Eng. C: Methods* **21** 46–51

Ko A C-T, Choo-Smith L-P, Hewko M, Sowa M G, Dong C C S and Cleghorn B 2006 Detection of early dental caries using polarized Raman spectroscopy *Opt. Express* **14** 203–15

Krafft C and Sergo V 2006 Biomedical applications of Raman and infrared spectroscopy to diagnose tissues *Spectroscopy* **20** 738186

Krishnamurithy G, Mohan S, Yahya N A, Mansor A, Murali M R, Balaji Raghavendran H R, Choudhary R, Sasikumar S and Kamarul T 2019 The physicochemical and biomechanical profile of for sterite and its osteogenic potential of mesenchymal stromal cells *PLoS One* **14** e0214212

Kudelski A 2008 Analytical applications of Raman spectroscopy *Talanta* **76** 1–8

Liang S, Wang M, Wang Y and Jiang T 2020 A pilot study about the effect of laser-induced fluorescence on color and translucency of human enamel during tooth bleaching *Photobiomodul. Photomed. Laser Surg.* **38** 151–9

Loguercio A D, Muñoz M A, Luque-Martinez I, Hass V, Reis A and Perdigão J 2015 Does active application of universal adhesives to enamel in self-etch mode improve their performance? *J. Dent.* **43** 1060–70

Lupi S M, Albini B, Baena A R Y, Lanfrè G and Galinetto P 2020a Anatase forming treatment without surface morphological alteration of dental implant *Materials* **13** 5280

Lupi S M, Galinetto P, Albini B, Di Ronza E, Rizzo S and y Baena R R 2020b Micro-Raman spectroscopy of dental implants subjected to different surface treatments *Appl. Sci.* **10** 2417

Lutin A, Bulatov V, Jadwat Y, Wood N H, Feller L and Schechter I 2015 Fourier transform spectral imaging microscopy (FT-SIM) and scanning Raman microscopy for the detection of indoor common contaminants on the surface of dental implants *Talanta* **134** 514–23

Mohn D, Zehnder M, Imfeld T and Stark W J 2010 Radio-opaque nanosized bioactive glass for potential root canal application: evaluation of radiopacity, bioactivity and alkaline capacity *Int. Endod. J.* **43** 210–7

Moonesi Rad R, Atila D, Evis Z, Keskin D and Tezcaner A 2019 Development of a novel functionally graded membrane containing boron-modified bioactive glass nanoparticles for guided bone regeneration *J. Tissue Eng. Regen. Med.* **13** 1331–45

Moreira M A, Souza N O, Sousa R S, Freitas D Q, Lemos M V, De Paula D M, Maia F J N, Lomonaco D, Mazzetto S E and Feitosa V P 2017 Efficacy of new natural biomodification agents from anacardiaceae extracts on dentin collagen cross-linking *Dental Mater.* **33** 1103–9

Moshaverinia A, Roohpour N, Billington R W, Darr J A and Rehman I U 2008 Synthesis of *N*-vinylpyrrolidone modified acrylic acid copolymer in supercritical fluids and its application in dental glass–ionomer cements *J. Mater. Sci.—Mater. Med.* **19** 2705–11

Movasaghi Z, Rehman S and Ur Rehman I 2008 Fourier transform infrared (FTIR) spectroscopy of biological tissues *Appl. Spectrosc. Rev.* **43** 134–79

Nafie L A 2020 Recent advances in linear and nonlinear Raman spectroscopy. Part XIV *J. Raman Spectrosc.* **51** 2354–76

Nakonieczny D, Paszenda Z, Drewniak S, Radko T and Lis M 2016 $SZrO_2$–CeO_2 ceramic powders obtained from a sol–gel process using acetylacetone as a chelating agent for potential application in prosthetic dentistry *Acta Bioeng. Biomechan.* **18** 53–60

Nascimento G M D 2018 The multiple applications of Raman spectroscopy *Raman Spectroscopy* (Rijeka: InTech)

Okulus Z, Buchwald T, Czarnecka B and Voelkel A 2017 The effect of bonding system application on surface characteristics of bovine dentin and enamel *Mater. Sci. Eng. C: Mater. Biol. Appl.* **76** 1224–31

Oliveira B, Ulbaldini A, Sato F, Baesso M L, Bento A C, Andrade L, Lima S M and Pascotto R C 2017 Chemical interaction analysis of an adhesive containing 10-methacryloyloxydecyl dihydrogen phosphate (10-MDP) with the dentin in noncarious cervical lesions *Oper. Dent.* **42** 357–66

Osorio R, Osorio E, Cabello I and Toledano M 2014 Zinc induces apatite and scholzite formation during dentin remineralization *Caries Res.* **48** 276–90

Osorio R, Sauro S, Watson T F and Toledano M 2016 Polyaspartic acid enhances dentine remineralization bonded with a zinc-doped Portland-based resin cement *Int. Endod. J.* **49** 874–83

Osorio R, Osorio E, Aguilera F S, Medina-Castillo A L, Toledano M and Toledano-Osorio M 2018 Silver improves collagen structure and stability at demineralized dentin: a dynamic-mechanical and Raman analysis *J. Dent.* **79** 61–7

Ou K-L, Hsu T-C, Liu Y-C, Yang K-H and Tsai H-Y 2014 Silver overlayer-modified surface-enhanced Raman scattering-active gold substrates for potential applications in trace detection of biochemical species *Anal. Chim. Acta* **806** 188–96

Padmanabhan V P, Kulandaivelu R, Narayanan S and Nellaiappan T S 2018 New core–shell hydroxyapatite/Gum-acacia nanocomposites for drug delivery and tissue engineering applications *Mater. Sci. Eng. C: Mater. Biol. Appl.* **92** 685–93

Pelaez-Vargas A, Gallego-Perez D, Magallanes-Perdomo M, Fernandes M H, Hansford D J, De Aza A H, Pena P and Monteiro F J 2011 Isotropic micropatterned silica coatings on zirconia induce guided cell growth for dental implants *Dental Mater.* **27** 581–9

Penkov O V, Pukha V E, Starikova S L, Khadem M, Starikov V V, Maleev M V and Kim D-E 2016 Highly wear-resistant and biocompatible carbon nanocomposite coatings for dental implants *Biomaterials* **102** 130–6

Perillo L, d'Apuzzo F, Illario M, Laino L, Di Spigna G, Lepore M and Camerlingo C 2020 Monitoring biochemical and structural changes in human periodontal ligaments during orthodontic treatment by means of micro-Raman spectroscopy *Sensors* **20** 497

Pongprueksa P, De Munck J, Duca R C, Poels K, Covaci A, Hoet P, Godderis L, Van Meerbeek B and Van Landuyt K L 2015 Monomer elution in relation to degree of conversion for different types of composite *J. Dent.* **43** 1448–55

Portella F F, Santos P D, Lima G B, Leitune V C B, Petzhold C L, Collares F M and Samuel S M W 2014 Synthesis and characterization of a glycerol salicylate resin for bioactive root canal sealers *Int. Endod. J.* **47** 339–45

Prajapati S, Tao J, Ruan Q, De Yoreo J J and Moradian-Oldak J 2016 Matrix metalloproteinase-20 mediates dental enamel biomineralization by preventing protein occlusion inside apatite crystals *Biomaterials* **75** 260–70

Raman C V and Krishnan K S 1928 A new type of secondary radiation *Nature* **121** 501–2

Ramos C M, Tabata A S, Cesar P F, Rubo J H, Silveira Fracisconi P A and Sanches Borges A F 2015 Application of micro-Raman spectroscopy to the study of yttria-stabilized tetragonal zirconia polycrystal (Y-TZP) phase transformation *Appl. Spectrosc.* **69** 810–4

Randolph L D, Palin W M, Bebelman S, Devaux J, Gallez B, Leloup G and Leprince J G 2014 Ultra-fast light-curing resin composite with increased conversion and reduced monomer elution *Dental Mater.* **30** 594–604

Ranjkesh B, Chevallier J, Salehi H, Cuisinier F, Isidor F and Løvschall H 2016 Apatite precipitation on a novel fast-setting calcium silicate cement containing fluoride *Acta Biomater. Odontol. Scand.* **2** 68–78

Redha O, Strange A, Maeva A, Sambrook R, Mordan N, McDonald A and Bozec L 2019 Impact of carbamide peroxide whitening agent on dentinal collagen *J. Dent. Res.* **98** 443–9

Ren N, Li J, Qiu J, Yan M, Liu H, Ji D, Huang J, Yu J and Liu H 2017 Growth and accelerated differentiation of mesenchymal stem cells on graphene-oxide-coated titanate with dexamethasone on surface of titanium implants *Dental Mater.* **33** 525–35

Rodrigues B V, Cs Leite N, das Neves Cavalcanti B, da Silva N S, Marciano F R, Corat E J, Webster T J and Lobo A O 2016 Graphene oxide/multi-walled carbon nanotubes as nanofeatured scaffolds for the assisted deposition of nanohydroxyapatite: characterization and biological evaluation *Int. J. Nanomed.* **11** 2569–85

Rohman A, Windarsih A, Lukitaningsih E, Rafi M, Betania K and Fadzillah N A 2020 The use of FTIR and Raman spectroscopy in combination with chemometrics for analysis of biomolecules in biomedical fluids: a review *Biomed. Spectrosc. Imaging.* **8** 55–71

Rokaya D, Srimaneepong V, Qin J, Siraleartmukul K and Siriwongrungson V 2019 Graphene oxide/silver nanoparticle coating produced by electrophoretic deposition improved the mechanical and tribological properties of NiTi alloy for biomedical applications *J. Nanosci. Nanotechnol.* **19** 3804–10

Sá Ferreira E M de, Soares L E S, Antunes H S, Uemura S T, Barbosa P da S, Salmon Jr H A and Sant'Anna G R de 2016 Effect of therapeutic doses of radiotherapy on the organic and inorganic contents of the deciduous enamel: an *in vitro* study *Clin. Oral Investigations* **20** 1953–61

Sağlam B C, Koçak M M, Koçak S, Türker S A and Arslan D 2015 Comparison of Nd:YAG and diode laser irradiation during intracoronal bleaching with sodium perborate: color and Raman spectroscopy analysis *Photomed. Laser Surg.* **33** 77–81

Santini A and Miletic V 2008 Quantitative micro-Raman assessment of dentine demineralization, adhesive penetration, and degree of conversion of three dentine bonding systems *Eur. J. Oral Sci.* **116** 177–83

Sauro S, Thompson I and Watson T F 2011 Effects of common dental materials used in preventive or operative dentistry on dentin permeability and remineralization *Oper. Dent.* **36** 222–30

Schlücker S 2014 Surface-enhanced Raman spectroscopy: concepts and chemical applications *Angew. Chem.—Int. Ed.* **53** 4756–95

Schofield S C, Berno B, Langman M, Hall G and Filiaggi M J 2006 Gelled calcium polyphosphate matrices delay antibiotic release *J. Dent. Res.* **85** 643–7

Sender B, Ricardo F, Castañeda Vía J A and Tay L Y 2020 Influence of different phosphoric acids before application of universal adhesive on the dental enamel *J. Esthetic Restor. Dent.* **32** 797–805

Sheikh Z, Samad Khan A, Roohpour N, Glogauer M and Rehman I U 2016 Protein adsorption capability on polyurethane and modified-polyurethane membrane for periodontal guided tissue regeneration applications *Mater. Sci. Eng. C: Mater. Biol. Appl.* **68** 267–75

Singh S P, Ibrahim O, Byrne H J, Mikkonen J W, Koistinen A P, Kullaa A M and Lyng F M 2016 Recent advances in optical diagnosis of oral cancers: review and future perspectives *Head. Neck* **38** E2403–11

Siqueira F S F de, Armas-Vega A, Izquierdo-Bucheli A, Pinto T F, Hanzen T A, Bauer J, Millan Cardenas A F and Loguercio A D 2019 Does the conditioning mode and duration of universal adhesives affect the bonding effectiveness to fluorotic enamel? *J. Adhesive Dent.* **21** 525–36

Soares L E S, Nahórny S, de Faria Braga V, Marciano F R, Bhattacharjee T T and Lobo A O 2020 Raman spectroscopy-multivariate analysis related to morphological surface features on nanomaterials applied for dentin coverage *Spectrochim. Acta.* A **228** 117818

Sorrentino R, Navarra C O, Di Lenarda R, Breschi L, Zarone F, Cadenaro M and Spagnuolo G 2019 Effects of finish line design and fatigue cyclic loading on phase transformation of zirconia dental ceramics: a qualitative micro-Raman spectroscopic analysis *Materials* **12** 863

Sousa C J A, Pereira M C, Almeida R J, Loyola A M, Silva A C A and Dantas N O 2014 Synthesis and characterization of zinc oxide nanocrystals and histologic evaluation of their biocompatibility by means of intraosseous implants *Int. Endod. J.* **47** 416–24

Spies B C, Zhang F, Wesemann C, Li M and Rosentritt M 2020 Reliability and aging behavior of three different zirconia grades used for monolithic four-unit fixed dental prostheses *Dental Mater.* **36** e329–39

Talari A C S, Garcia Martinez M A, Movasaghi Z, Rehman S and Ur Rehman I 2017 Advances in Fourier transform infrared (FTIR) spectroscopy of biological tissues *Appl. Spectrosc. Rev.* **52** 456–506

Talungchit S, Jessop J L P, Cobb D S, Qian F, Geraldeli S, Pashley D H and Armstrong S R 2014 Ethanol-wet bonding and chlorhexidine improve resin–dentin bond durability: quantitative analysis using Raman spectroscopy *J. Adhesive Dent.* **16** 441–50

Tetschke F, Kirsten L, Golde J, Walther J, Galli R, Koch E and Hannig C 2018 Application of optical and spectroscopic technologies for the characterization of carious lesions *in vitro Biomedizinische Technik. Biomed. Eng.* **63** 595–602

Toledano-Osorio M, Cabello I, Lynch C D and Aguilera F S 2018 Mild acids facilitate functional dentin remineralization under thermo-mechanical stimuli *Am. J. Dent.* **31** 155–65

Toledano M, Pérez-Álvarez M C, Aguilera F S, Osorio E, Cabello I, Toledano-Osorio M and Osorio R 2017 A zinc oxide-modified hydroxyapatite-based cement facilitated new crystalline-stoichiometric and amorphous apatite precipitation on dentine *Int. Endod. J.* **50** e109–19

Toledano M, Aguilera F S, Cabello I and Osorio R 2014a Masticatory function induced changes, at subnanostructural level, in proteins and mineral at the resin–dentine interface *J. Mech. Behav. Biomed. Mater.* **39** 197–209

Toledano M, Aguilera F S, Osorio E, Cabello I and Osorio R 2014b Microanalysis of thermal-induced changes at the resin–dentin interface *Microsc. Microanal.* **20** 1218–33

Toledano M, Aguilera F S, Sauro S, Cabello I, Osorio E and Osorio R 2014c Load cycling enhances bioactivity at the resin–dentin interface *Dental Mater.* **30** e169–88

Toledano M, Cabello I, Aguilera F S, Osorio E and Osorio R 2015 Effect of *in vitro* chewing and bruxism events on remineralization, at the resin–dentin interface *J. Biomech.* **48** 14–21

Toledano M, Cabello I, Cabrerizo Vílchez M A, Fernández M A and Osorio R 2014d Surface microanalysis and chemical imaging of early dentin remineralization *Microsc. Microanal.* **20** 245–56

Toledano M, Osorio E, Aguilera F S, Sauro S, Cabello I and Osorio R 2014e *In vitro* mechanical stimulation promoted remineralization at the resin/dentin interface *J. Mech. Behav. Biomed. Mater.* **30** 61–74

Tsuda H and Arends J 1997 Raman spectroscopy in dental research: a short review of recent studies *Adv. Dental Res.* **11** 539–47

Uslu O, Dikmen G and Orhan E O 2021 Intrinsic physicochemical interactions of calcium hydroxide-based medications *Microsc. Res. Tech.* **84** 432–40

Vandenabeele P and Edwards H (ed) 2018Raman Spectroscopy in Archaeology and Art History vol 2 (London: The Royal Society of Chemistry)

Vieira K A, Steiner-Oliveira C, Soares L E S, Rodrigues L K A and Nobre-dos-Santos M 2015 *In vitro* evaluation of enamel demineralization after several overlapping CO_2 laser applications *Lasers Med. Sci.* **30** 901–7

Wang Y, Song Y, Tao Y, Muhamadali H, Goodacre R, Zhou N-Y, Preston G M, Xu J and Huang W E 2016 Reverse and multiple stable isotope probing to study bacterial metabolism and interactions at the single cell level *Anal. Chem.* **88** 9443–50

Yakubu E, Li B, Duan Y and Yang S 2018 Full-scale Raman imaging for dental caries detection *Biomed. Opt. Express* **9** 6009–16

Yao L, Peng C and Wu J 2013 Wettability and bond strength between leucite-reinforced dental porcelains and Co–Cr alloy *J. Prosthet. Dent.* **110** 515–20

Zhang F *et al* 2020 Reliability of an injection-moulded two-piece zirconia implant with PEKK abutment after long-term thermo-mechanical loading *J. Mech. Behav. Biomed. Mater.* **110** 103967

Zhang J, Boyes V, Festy F, Lynch R J M, Watson T F and Banerjee A 2018 *In-vitro* subsurface remineralisation of artificial enamel white spot lesions pre-treated with chitosan *Dental Mater.* **34** 1154–67

Zhao L, Kim T-H, Kim H-W, Ahn J-C and Kim S Y 2015 Surface-enhanced Raman scattering (SERS)-active gold nanochains for multiplex detection and photodynamic therapy of cancer *Acta Biomater.* **20** 155–64

Chapter 3

Applications of fluorescence spectroscopy in dentistry

Monalisa Mishra

Teeth and physiology are intertwined with each other since teeth reside in the oral cavity, the opening of the digestive system. The role of the gut–brain axis in many neurodegenerative diseases directs the need for more research into dental-associated disease. Thus the identification of abnormalities and the proper treatment of teeth will help to prevent the onset of life-threatening neurodegenerative disease. Several conventional and unconventional methods are used to detect the defects associated with teeth. Among all these methods fluorescence has the upper hand since it is a non-invasive technique. Different parts of the teeth have their own autofluorescence. The autofluorescence properties of teeth change with age, infection and the minerals accumulated over it. This fluorescence property of the teeth is used to detect carious lesions. Several set-ups and devices have been developed which use fluorescence to detect carious lesions of the teeth. Since it is a non-invasive technique it can be used in children several times to detect carious lesions in the planktonic as well as in the biofilm stage. Using this technique the status of restorative dental materials can also be checked. The current chapter summarises the application of fluorescence spectroscopy in several dental applications.

3.1 Introduction

Although the phenomenon of fluorescence was discovered in 1845, its occurrence in rabbit teeth was identified for the first time by Benedict in 1928 [1]. He detected that dentine fluoresce was brighter than that of enamel. However, he could not detect this phenomenon in teeth with white lesions. After this observation he proposed for the first time that fluorescence could be used to identify dental caries [1, 2]. He also proposed that the demineralised area of the teeth can also be visualised by enhancing the contrast. When a tooth is exposed to a certain wavelength of light under UV light it emits in the visible range [3]. Tooth fluorescence was defined as an optical

doi:10.1088/978-0-7503-3459-4ch3

property that emits in a visible range of 430–450 nm. Later, this phenomenon was used to diagnose many carious lesions of the teeth. A few decades later it was found that organic matter such as protein present within the teeth contributes to the teeth fluorescence [4–6]. Tooth enamel is approximately 96% mineral and 1%–2% organic material. Human dentine has a 30%–40% organic matrix [7]. The presence of a greater amount of organic matrix contributes to the strong fluorescence of the dentine of the teeth.

The presence of the fluorescence properties of teeth allows the health of teeth to be detected using fluorescence spectroscopy. It is a noncontact, non-destructive and sensitive procedure with high specificity [8]. Consequently, it has the upper hand over other imaging techniques [9]. In addition, autofluorescence, UV irradiation, the black lights used in nightclubs and UV coated lamps can induce fluorescence in teeth. The first clinical tooth fluorometer was developed by considering the fluorescence properties of teeth [10]. Later, fluorometers were used to detect the fluorescence properties of the restorative materials used in dentistry [11–13]. A fluorescence set-up to detect tooth defects requires a UV light, a dark room and a DSLR camera [14]. This chapter aims to summarise the fluorescence properties of the teeth and its restorative materials in the detection and treatment of tooth associated defects.

3.2 The structure of human teeth

Human teeth are heterodont in nature and their structure, shape and size is regulated by several genetic factors. Although the development of teeth is a complex phenomenon, for all teeth it begins from calcified tissue which is enriched with hydroxyapatite and proteins [15]. The different regions of the teeth include the enamel, dentine, pulp, cementum and periodontal ligament (figure 3.1). The outermost part of teeth is known as enamel. It is the stiffest part of a tooth and is a few micrometres in length and few nanometres in width. It is enriched with calcium phosphate which is needle-shaped and has an ordered arrangement [16]. It is formed

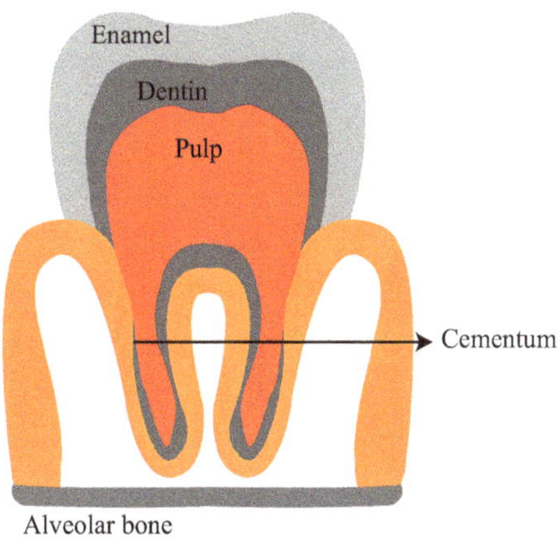

Figure 3.1. The different layers present within the teeth.

of nonliving tissues. The enamel does not undergo any change after it is formed. Just below the enamel is the dentine. It is the next hard structure after the enamel and it is a living tissue. It is formed of odontoblasts, the precursor cells of dentine [17]. Dentine consists of many microscopic tubules. The apatite cells are smaller and less organised in this region. Pulp is the soft part and is present in the inner part of the tooth. The blood vessels and the nerves run through it. The cementum is the region that binds the root of the tooth to the gum and jawbone. It prevents the root from being attacked by micro-organisms.

3.3 The development of human teeth

Humans develop two different types of teeth. One is primary teeth and the other is permanent teeth. Both types of teeth follow the same stages during development. The process of tooth development is called odontogenesis. It includes six different developmental stages: (i) dental epithelium, (ii) bud stage, (iii) cap stage, (iv) bell stage, (v) crown or root formation and (vi) eruption stage (figure 3.2). Primary tooth development occurs at eight weeks of development inside the uterus. The formation of dental epithelium begins from the dental lamina, which is a thick band of cells [18]. These cells form the tooth germ which is comprised of soft tissue and essential for tissue growth. This step is known as the bud stage. This leads to the cap stage. In this stage outer layer of the tooth is formed and covers the bud stage. The rest of the tooth bud is known as the dental papilla [18]. It forms the inner layer of the tooth, i.e. the dentine and pulp. Both the enamel and dental papilla are surrounded by the dental

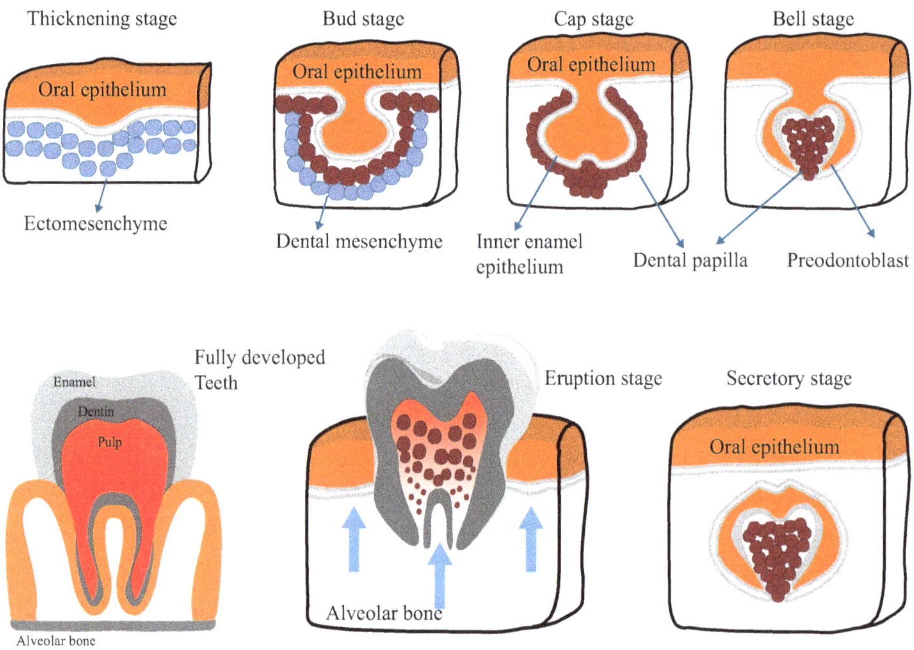

Figure 3.2. Stages involved in the development of teeth. Redrawn after Viale-Bouroncle *et al* [18].

follicle. The dental follicle is enriched with blood vessels and nerves. The cap stage results the formation of the three different layers of the tooth, i.e. the enamel, papilla and dental follicle [19]. This stage leads to the formation of the bell stage. At this stage the enamel grows into a bell stage and two key events take place. The first event includes the differentiation of the enamel organ. Depending on the function it has four different groups of cells: (i) the inner enamel epithelium, (ii) outer enamel epithelium, (iii) stratum intermedium and (iv) stellate reticulum. In the next stage the enamel epithelium folds to form the tooth crown and the dental lamina disintegrates. The fifth step of development includes the formation of the crown and root. The crown stage results in the formation of the enamel and dentine. Amyloblast cells are responsible for the formation of enamel and odontoblast cells give rise to dentine, corresponding to the enamel epithelium and dental papilla, respectively. The root is formed of root dentine and the root canal. The root has three different types of cells: (i) dental papilla, (ii) dental follicles and (iii) Hertwig's epithelial root sheath. The formation of the crown results in the teeth developing vertically within the oral cavity [20]. During eruption the jawbone resorbs and connective tissue breaks down which help to move the tooth outward. The primary teeth erupt after around six months of development [21].

3.4 Factors affecting the fluorescence properties of teeth

Fluorescence phenomena are described as the de-excitation of an excited molecule from the lowermost excited state to the ground state. Like many biological structures, teeth also possess autofluorescence. The fluorescence properties of teeth can be determined by a filter fluorometer [22]. Tooth fluorescence varies by temperature, type, age and other factors. The fluorescence is measured generally from the mandibular and maxillary arches [23]. The different types of teeth, such as the incisors, canines, premolars and molars, do not vary in terms of fluorescence property [24]. The autofluorescence of teeth is due to the fluorophores present in the enamel and dentine [22]. The enamel contributes less to fluorescence than the dentine [25]. The organic material present in the dentine gives the fluorescence property to the teeth. As a consequence, if the chroma of dentine increases the fluorescence decreases [26]. In the presence of dental caries a white spot appears and such areas do not fluoresce. Salivary calculus fluoresces orange in colour [27]. Human dentine fluoresces more than enamel. Although the fluorophore present in teeth is not yet known, the calcium apatite matrix present in the teeth helps in the emission of light [28]. Proteins and the inorganic matrix present in the teeth contribute to the fluorescence properties of teeth. The fluorescence properties share similarities with tryptophan, tyrosine and other fluorophore groups present within the teeth [29]. The emission and excitation wavelengths of the enamel and organic mineral have been determined [30]. The fluorescence of human dentine has been investigated using time-resolved fluorescence microscopy. Different parts of a tooth, such as enamel, dentine and pulp, undergo noticeable changes with age. Changes to the structure of enamel and dentine can change the colouration of the teeth [31] by altering the path of the light that passes through the teeth [32]. People of the same age have similar spectral profiles irrespective of sex. The optical properties of the teeth remain the same up to an age of 26.5 years and diminution occurs afterward [11]. Human teeth present two different types of fluorescence properties. One is the emission of fluorescence by ambient light and the other is fluorescence emission by a specific wavelength of artificial light, such a laser. The second method is used to detect dental caries and the aesthetic restorative materials associated with teeth [33]. Hydrolysate of dentine protein has similar optical behaviour to tryptophan, tyrosine and hydroxypyridiene [34]. Tryptophan, tyrosine and hydroxypyridiene fluoresce between 350 and 400 nm, but the fluorescence of dentine is at 440 nm [35]. The fluorescence property of dentine is thought to be due to the presence of the hydroxyapatite–pyridinoline complex. Spitzer and Bosch proved that the fluorescence property is due to the emitted wavelength of light, not due to the hydroxyapatite present within it. Currently, the fluorescence properties of teeth are due to the combined fluorescence of pyrimidine, tryptophan, tyrosine, hydroxypyridine, pyridinoline and dityrosine [36]. Mineralisation decreases the fluorescence intensity of the teeth and thus dentine fluoresces more than enamel. Similarly, with aging more mineralisation occurs in the teeth and as a result they fluoresce less. Hydrogen peroxide removes the organic content from dentine and thus decreases the fluorescence properties of teeth [37].

Growth of calculus over the teeth leads to the development of periodontitis. Hence, the removal of calculus is an alternative approach to the treatment of periodontitis. However, it is challenging to eliminate calculus by looking at a radiograph, or using visual and ultrasonic resonance methods [38–41]. The autofluorescence of supra- and subgingival calculus was determined efficiently by fluorescence spectroscopy [42]. Tryptophan autofluorescence was considered for this [43]. The fluorescence spectra depend on the colour of the calculus. The fluorescence spectra of unaffected teeth are in the violet–green region. The presence of micro-organisms in the calculus results in the presence of porphyrins in the teeth [44–46]. Dental deposits, bacteria and debris interfere with the fluorescent signal from the teeth.

3.5 Fluorescence properties needed for dental restorative materials

The detection of the fluorescence properties of teeth lead the development of fluorescence emitting dental restorative materials. Opaque dentine and translucent enamel were used to detect the translucent property of the teeth [47–49]. This lead to the development of restorative materials which have the same fluorescence properties as natural teeth [50, 51]. It is ideal for a restorative material to match the natural environment. Resin is the main component of restorative material and does not have fluorescence properties. To make it fluorescent, europium, cerium and ytterbium are added to it [52]. The composite material appears to be the same colour during daylight but it changes its colour when visualised under a fluorescent lamp. Changing UV lamps also changes the colour of the resin composite [53]. Dental restorative materials have the same fluorescence properties as natural teeth and this allows them to be used for epidemiological, forensic and clinical practice. Although the resins developed have both fluorescence and nonfluorescence properties, fluorescent materials have advantages over nonfluorescent ones [54]. The fluorescence spectra of dental resin have their emission in the blue range when exposed to UV light [55]. A resin composite of dentine affects the translucence properties of teeth [13]. Human dentine has its fluorescence peak at 440–450 nm. Five commercially available resins also show fluorescence in the same range. The level of the UV component of daylight affects the fluorescence properties of the resin [56]. A tooth and porcelain can be easily distinguished based on their fluorescence properties. Fluorescence property of silicate cement, resin and acrylic resin were detected using fluorescence technique. Silicate cement shows intensity at 430 and 525 nm [22]. The final fluorescence properties of restorative material arise from the combined action of all the materials present within the teeth. The fluorescence properties of composite resin also change with temperature, aging and bleaching agents [57–59].

3.6 Fluorescence properties of oral bacteria

An imbalance of bacteria occurs during oral or dental infection. As a result, it is essential is to detect the presence of bacteria in an infectious mouth. Although there are conventional methods such as microscopy techniques to detect different types of bacteria, fluorescence spectroscopy plays a unique role in detecting the types of bacteria. Many bacteria have autofluorescence and this property is often used for their classification [60, 61]. The fluorescence properties of *Staphylococcus aureus* and *Clostridium difficile* have been detected in clinical studies [62]. Lennon *et al* reported the different fluorescence properties of 11 different bacteria [63]. The biofilm properties of dental plaque vary depending on the level of sugar and pH in the surrounding environment [64–66]. *Streptococcus mutans*, *Prevotella intermedia* and *Candida albicans* exhibited different autofluorescence in the presence of specific light [67]. The autofluorescence of different oral bacteria was examined by Kang *et al* [68]. A fluorescence peak at 440 nm is due to the presence of NAD(P)H [69]. A peak present at 640 nm corresponds to the presence of the porphyrin molecule [70]. *Fusobacterium nucleatum* and *Lactobacillus gasseri* have a peak at 520 nm which corresponds to the flavin molecule [71]. *Actinomyces israelii* have a strong peak at 610–650 nm [68]. *P. gingivalis* have a dark colony and weak fluorescence properties. The presence of metal ions or hemin in the culture medium often hampers the fluorescence properties [64]. More research is needed to identify the fluorescence properties of bacteria in detecting dental disease.

3.7 Fluorescence as a diagnostic tool in dentistry

For caries detection, wavelength-dependent fluorescence properties were used. The fluorescence spectra of a normal tooth are dependent on the wavelength it is exciting [72]. Caries on hard tissue fluoresce whereas soft tissue does not [73]. Different caries fluoresce at different wavelengths, such as 488, 407, 400 to 420, 337 and 405, 442, 532 nm [74]. The fluorescence spectra of enamel and dentine at 337, 488, 515 and 613 nm were measured [75]. Enamel with carious lesions has lower fluorescence intensity and red-shift. The bacteria or denatured breakdown product present in teeth might interfere with the fluorescence properties of the teeth [22]. Once the microhardness of the dentine is below 25 Knoop hardness then the fluorescence signal increases. The spectral features of various caries have been studied [76]. Healthy tissues lack any emission band in the red region. In carious lesions, typical spectra are found in the fluorescent porphyrin and protoporphyrin IX region [45]. For a white spot, lesion red-shift is at minimum and for a dark brown spot it is at maximum. For porphyrin, a strong band at 600–700 nm was detected. The caries formation *in vitro* was measured using a 337 nm laser [77]. For normal teeth, the intensive maxima are at 480–500 nm and 430–450 nm [78]. During the formation of caries, maxima appear at 590–650 nm. The fluorescence spectra of human dentine, enamel and whole teeth were detected at 405, 444 and 532 nm. With the increase of excitation wavelength the emission peak shifts towards a longer wavelength. Brokos *et al* developed a photography technique using macro-flash UV-A 365 nm illumination [79].

3.8 Fluorescence-based spectral devices to detect carious teeth

The fluorescence emission of dental calculus varies with the instrument used to detect the fluorescence [22]. Autofluorescence is a valuable tool to detect subgingival calculus and dental caries [80]. Quantitative Light-induced Fluorescence (QLF; Inspektor Dental Care, Netherlands) measures wavelengths from 290 nm to 450 nm. At this wavelength the loss of minerals occurring in both enamel and dentine can be measured. Normal teeth fluoresce and carious tooth structures appear as black [81]. QLF can detect the defective area more efficiently than other caries detection methods. It can differentiate between the presence and absence of fluoride-containing regions of the teeth *in vivo* [82]. It detects qualitative light-induced fluorescence (QLF) parameters such as area (a, mm^2), fluorescence loss (F, %), and Q ($\%X$, mm^2) [83]. ΔQ and ΔF determine the mineral loss of the teeth [84]. DIAGNOdent (DD; KavoDEntal GmbH, Germany) is another device that uses a laser to excite pigments present in the carious teeth [22]. It uses the wavelength of 655 nm to detect the signal from bacteria and their by-products. It can detect a fluorescence intensity between 1 and 100 [27]. Mouth bacteria are strongly fluorescent, either in the single form or in the planktonic form. The intensity of the bacteria can be determined at the picomolar level [27]. The enamel is more transparent in the near-infrared region. Thus using this device, lesions in the dentine–enamel region can be detected with the highest accuracy, which is not possible in radiography techniques [9]. However, DD can detect only the bacteria present within the teeth, not the minerals of the teeth. White spots in the teeth are due to the presence of porphyrin within the tooth surface. DD can also detect the white lesions associated with the teeth. DD could detect the enamel and dentine of carious lesions in 32 different studies. However, DD is not efficient enough to detect caries under a dental sealer. DD technology is available in three different forms: (i) DD classic, (ii) DD Pen and (iii) KEY-3 Laser. Both DD Classic and DD Pen have a common basic principle but the latter is more sensitive. DD Pen can detect more bacteria within the cavity of the teeth. Occlusal caries were detected using this technique. Anatomical factors of the teeth interfere while detecting occlusal caries [85]. To remove this fluorescence from the teeth surgical excavation was needed [86]. By using this instrument the demineralisation of the dentine was evaluated [87]. Using a SporoLife® intra-oral camera the fluorescence intensity of normal and carious teeth can be distinguished. Since very few studies have been conducted in this area, more work is needed to validate this camera for occlusion caries lesions. A two-photon time-correlated fluorescence technique is used to detect the autofluorescence of subgingival calculus [88]. Er:YAG laser debridement along with an automatic calculus detector improves the calculus detection of the device [89]. An LED-based optical probe can detect subgingival calculus [90]. Human teeth illuminated by blue LED light at 405 nm is used to confirm clinical and histological findings. A fluorescence-based real-time detection is used to quantify the dental calculus [8].

3.9 Advantages of DIAGNOdent over other systems

DD is a small laser device which can scan the teeth for very preliminary decay. It is a pen-like structure which can glide over the teeth smoothly [91]. It can differentiate control teeth from infected teeth. It uses laser fluorescence to detect caries. This device is comprised mainly of a control unit and a hand-held probe. They are connected to each other by means of a cable with principal and decedent optical glass fibres. The probe has two attachments: (i) an occlusal surface and (ii) a flat tip or smooth (figure 3.3). DD uses a laser light of 655 nm since the control teeth do not fluoresce at this wavelength [92]. The detector is at 680 nm. Laser light is transmitted through the tip of the optical fibre and comes into contact with the tooth material. The emitted fluorescence and the backscattered light are collected by the tip.

The enamel and dentine have some autofluorescence due to presence of chromophores. Red and infrared fluorescence are less absorbed and scattered by the enamel than light of shorter wavelengths. This allows carious dentine to be measured. Carious dentine fluoresces differently due to the presence of bacterial biofilm. This is attributed to the presence of porphyrin, a degraded product of bacteria. The fluorescence of red light can differentiate between a carious and a normal tooth. The DD device can detect a value between 1 and 100 [93]. A value between 0 and 14 is considered as no caries. A value between 15 and 20 needs attention and is confined only to the enamel. A value between 20 and 100 is suggestive of serious decay.

DD can detect caries with an accuracy of 90%. Using a conventional approach the probe cannot reach the carious lesion with precision. In this regard the laser light of DD can penetrate the enamel and detect a carious lesion. Thus DD helps in

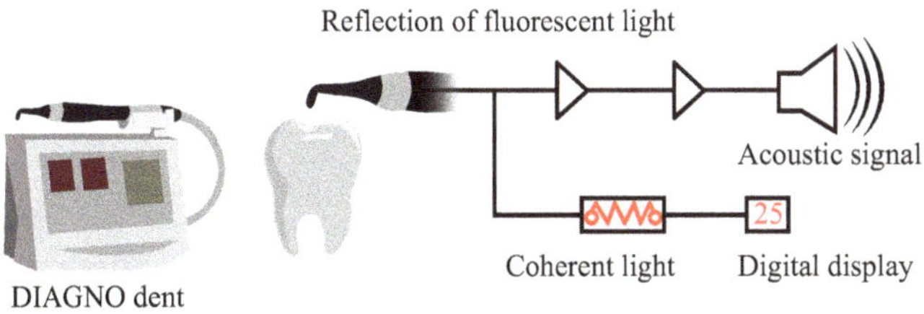

Figure 3.3. The set-up of a DIAGNOdent device [91, 92].

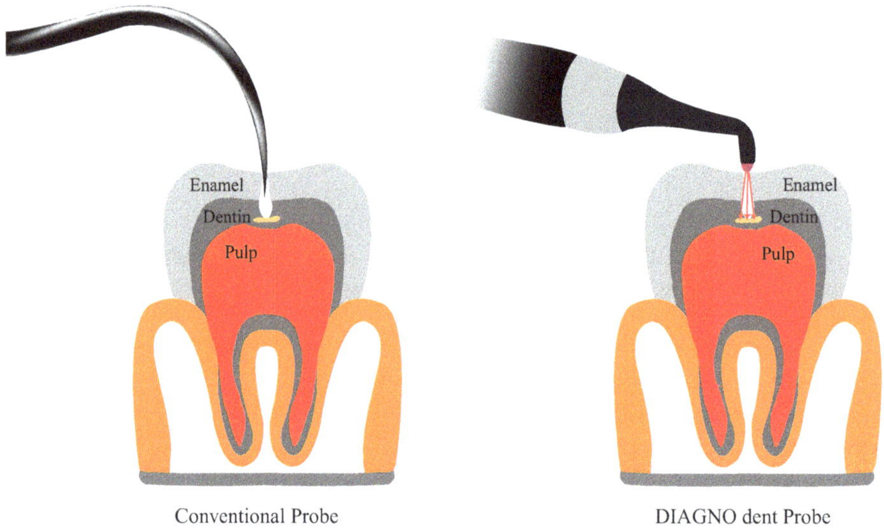

Figure 3.4. Difference between the caries detection of a DIAGNOdent probe and a conventional probe.

detecting caries in a more accurate way than the conventional method [94] (figure 3.4).

3.10 Conclusion

In most cases, the carious lesions of children are detected using radiography techniques. The application of fluorescence as an adjunct technique will improve the reproducibility of the detection of carious lesions. Moreover, fluorescence uses non-ionising radiation, hence it is safe to apply it multiple times. The deposition of plaque and calculus show fluorescence. In such cases cleaning of the tooth surface before examination gives more accurate results. The use of ultra-blue light and an intra-oral camera help to detect the spots present in the teeth efficiently. The detection of caries at an early stage and remineralization of the teeth helps to arrest and reverse caries. Several fluorescent dyes are added to detect the demineralisation of teeth. The use of a multispectral fluorescence technique will allow the detection of the fluorescence of different parts of the teeth, caries, and restorative materials. The use of the measurement of the backscattering of light along with fluorescence will be a more accurate way to detect the defects associated with teeth. Since fluorescence uses low-power light it can be used as an indispensable tool for dentistry in the future.

Acknowledgements

MM's lab is supported by financial assistance from the Department of Biotechnology, Ministry of Science and Technology (SERB/EMR/2017/003054, BT/PR21857/NNT/28/1238/2017, and Odisha DBT 3325/ST(BIO)-02/2017).

References

[1] Benedict H 1928 A note on the fluorescence of teeth in ultra-violet rays *Science* **67** 442

[2] Benedict H 1929 The fluorescence of teeth as another method of attack on the problem of dental caries *J. Dent. Res.* **9** 274–5

[3] Spitzer D and Ten J B 1976 The total luminescence of bovine and human dental enamel *Calcif. Tissue Res.* **20** 201–8

[4] Armstrong W and Singer L 1963 Fluoride contents of enamel of sound and carious human teeth: a reinvestigation *J. Dent. Res.* **42** 133–6

[5] Armstrong W 1963 Fluorescence characteristics of sound and carious human dentine preparations *Arch. Oral. Biol.* **8** 79–90

[6] Shore V and Pardee A B 1956 Fluorescence of some proteins, nucleic acids and related compounds *Arch. Biochem. Biophys.* **60** 100–7

[7] Mann A and Dickinson M 2006 Nanomechanics, chemistry and structure at the enamel surface *Monogr. Oral. Sci.* **19** 105

[8] Qin Y, Luan X, Bi L, Lü Z, Sheng Y, Somesfalean G, Zhou C and Zhang Z 2007 Real-time detection of dental calculus by blue-LED-induced fluorescence spectroscopy *J. Photochem. Photobiol.* B **87** 88–94

[9] Satpathy A, Ranjan R, Priyadarsini S, Gupta S, Mathur P and Mishra M 2019 *Medical Imaging Methods* (Berlin: Springer) pp 59–95

[10] Hall J, Hefferren J and Olsen N 1970 Study of fluorescent characteristics of extracted human teeth by use of a clinical fluorometer *J. Dent. Res.* **49** 1431–6

[11] Da Silva R D, da Silva M A D, de Oliveira O B, Melo A C M and de Oliveira R N 2013 Dental fluorescence: potential forensic use *Forensic Sci. Int.* **231** 167–71

[12] Kvaal S and Solheim T 1989 Fluorescence from dentin and cementum in human mandibular second premolars and its relation to age *Eur. J. Oral Sci.* **97** 131–8

[13] Lee Y-K 2015 Translucency of human teeth and dental restorative materials and its clinical relevance *J. Biomed. Opt.* **20** 045002

[14] Brokos Y, Stavridakis M and Krejci I 2018 A novel method of capturing fluorescence in clinical dentistry *Compend. Contin. Educ. Dent.* **39** e1–4

[15] Busch S, Schwarz U and Kniep R 2001 Morphogenesis and structure of human teeth in relation to biomimetically grown fluorapatite–gelatine composites *Chem. Mater.* **13** 3260–71

[16] Priyadarsini S, Mukherjee S and Mishra M 2018 Nanoparticles used in dentistry: a review *J. Oral Biol. Craniofac. Res.* **8** 58–67

[17] Butler W T 1998 Dentin matrix proteins *Eur. J. Oral. Sci.* **106** 204–10

[18] Viale-Bouroncle S, Felthaus O, Schmalz G, Brockhoff G, Reichert T E and Morsczeck C 2012 The transcription factor DLX3 regulates the osteogenic differentiation of human dental follicle precursor cells *Stem Cell Dev.* **21** 1936–47

[19] McCollum M and Sharpe P T 2001 Evolution and development of teeth *J. Anat.* **199** 153–9

[20] Soukup V, Epperlein H-H, Horácek I and Cerny R 2008 Dual epithelial origin of vertebrate oral teeth *Nature* **455** 795–8

[21] Cunha R F, Boer F A C, Torriani D D and Frossard W T G 2001 Natal and neonatal teeth: review of the literature *Pediatr. Dent.* **23** 158–62

[22] Lee Y-K 2015 Fluorescence properties of human teeth and dental calculus for clinical applications *J. Biomed. Opt.* **20** 040901

[23] Haworth K E, Healy C, Morgan P and Sharpe P T 2004 Regionalisation of early head ectoderm is regulated by endoderm and prepatterns the orofacial epithelium *Development* **131** 4797–806

[24] Matsumoto H, Kitamura S and Araki T 1999 Autofluorescence in human dentine in relation to age, tooth type and temperature measured by nanosecond time-resolved fluorescence microscopy *Arch. Oral. Biol.* **44** 309–18

[25] Rousseau C, Vaidya S, Creanor S, Hall A, Girkin J, Whitters C, Strang R and McHugh S 2002 The effect of dentine on fluorescence measurements of enamel lesions *in vitro Caries Res.* **36** 381–5

[26] Caneppele T, Borges A B and Torres C 2013 Effects of dental bleaching on the color, translucency and fluorescence properties of enamel and dentin *Eur. J. Esthet. Dent.* **8** 200–12

[27] Walsh L J 2018 Caries diagnosis aided by fluorescence *Dental Caries—Diagnosis and Management* ed Z Arkanslan (Rijeka: InTech) pp 97–115

[28] Zhang L, Nelson L Y and Seibel E J 2011 Red-shifted fluorescence of sound dental hard tissue *J. Biomed. Opt.* **16** 071411

[29] Berezin M Y and Achilefu S 2010 Fluorescence lifetime measurements and biological imaging *Chem. Rev.* **110** 2641–84

[30] Lee K, Kim H, Kim K and Kwon Y 2006 Mineral loss from bovine enamel by a 30% hydrogen peroxide solution *J. Oral. Rehabil.* **33** 229–33

[31] Wongkhantee S, Patanapiradej V, Maneenut C and Tantbirojn D 2006 Effect of acidic food and drinks on surface hardness of enamel, dentine, and tooth-coloured filling materials *J. Dent.* **34** 214–20

[32] Carey C M 2014 Tooth whitening: what we now know *J. Evid.-Based Dent. Pract.* **14** 70–6

[33] Balhaddad A A, Kansara A A, Hidan D, Weir M D, Xu H H and Melo M A S 2019 Toward dental caries: exploring nanoparticle-based platforms and calcium phosphate compounds for dental restorative materials *Bioact. Mater.* **4** 43–55

[34] Hoerman K and Mancewicz S A 1964 Phosphorescence of calcified tissues *Arch. Oral. Biol.* **9** 517–IN3

[35] Fukushima Y 1987 Topography of fluorescence and its possible composites in human teeth *Cell. Mol. Biol.* **33** 725–36

[36] Volpato C A M, Pereira M R C and Silva F S 2018 Fluorescence of natural teeth and restorative materials, methods for analysis and quantification: a literature review *J. Esthet. Restor. Dent.* **30** 397–407

[37] Jiang T, Ma X, Wang Y, Tong H, Shen X, Hu Y and Hu J 2008 Investigation of the effects of 30% hydrogen peroxide on human tooth enamel by Raman scattering and laser-induced fluorescence *J. Biomed. Opt.* **13** 014019

[38] Sherman P, Hutchens Jr L, Jewson L, Moriarty J, Greco G and McFall Jr W 1990 The effectiveness of subgingival scaling and root planing I. Clinical detection of residual calculus *J. Periodontol.* **61** 3–8

[39] Rees J, Addy M and Hughes J 1999 An *in vitro* assessment of the dentine lost during instrumentation using the Periosonic® system *J. Clin. Periodontol.* **26** 106–9

[40] Tugnait A, Clerehugh V and Hirschmann P 2000 The usefulness of radiographs in diagnosis and management of periodontal diseases: a review *J. Dent.* **28** 219–26

[41] Kocher T, Langenbeck M, Rühling A and Plagmann H C 2000 Subgingival polishing with a Teflon-coated sonic scaler insert in comparison to conventional instruments as assessed on extracted teeth: (I) Residual deposits *J. Clin. Periodontol.* **27** 243–9

[42] Bürmen M, Fidler A, Pernuš F and Likar B 2012 Hyperspectral laser-induced autofluorescence imaging of dental caries *Proc. SPIE* **8208** 82080E
[43] Foreman P 1980 The excitation and emission spectra of fluorescent components of human dentine *Arch. Oral. Biol.* **25** 641–7
[44] Slimani A, Nouioua F, Panayotov I, Giraudeau N, Chiaki K, Shinji Y, Cloitre T, Levallois B, Gergely C and Cuisinier F 2016 Porphyrin and pentosidine involvement in the red fluorescence of enamel and dentin caries *Int. J. Exp. Dental Sci.* **5** 1–10
[45] Konigm K, Flelviming G and Hibst R 1998 Laser-induced autofluorescen CE spectroscopy of dental caries *Cell. Mol. Biol.* **44** 1293–300
[46] Bakhmutov D, Gonchukov S and Sukhinina A 2010 Fluorescence spectroscopy of dental calculus *Laser Phys. Lett.* **7** 384–7
[47] Dietschi D 1995 Free-hand composite resin restorations: a key to anterior aesthetics *Pract. Periodontics Aesthet. Dent.* **PPAD7** 15–25
[48] Dietschi D 2001 Layering concepts in anterior composite restorations *J. Adhes. Dent.* **3** 71–80
[49] Dietschi D 2000 Exploring the layering concepts for anterior teeth *Adhesion: The Silent Revolution In Dentistry* ed J F Rouiet and M Degrange (Tokyo: Quintessence Publishing Company) pp 235–51
[50] Joiner A 2004 Tooth colour: a review of the literature *J. Dent.* **32** 3–12
[51] Fondriest J 2003 Shade matching in restorative dentistry: the science and strategies *Int. J. Periodontics Restor. Dent.* **23** 467–80
[52] Uo M, Okamoto M, Watari F, Tani K, Morita M and Shintani A 2005 Rare earth oxide-containing fluorescent glass filler for composite resin *Dent. Mater. J.* **24** 49–52
[53] Lee Y-K, Kim J-H and Ahn J-S 2007 Influence of the changes in the UV component of illumination on the color of composite resins *J. Prosthet. Dent.* **97** 375–80
[54] Meller C and Klein C 2012 Fluorescence properties of commercial composite resin restorative materials in dentistry *Dent. Mater. J.* **31** 916–23
[55] Paulo H D, Pereira J C, Svizero N R, Rueggeberg F A and Pashley D H 2006 Use of fluorescent compounds in assessing bonded resin-based restorations: a literature review *J. Dent.* **34** 623–34
[56] Lee Y K, Lu H and Powers J M 2005 Fluorescence of layered resin composites *J. Esthet. Restor. Dent.* **17** 93–100
[57] Bueno R P, Salomone P, Villetti M A and Pozzobon R T 2013 Effect of bleaching agents on the fluorescence of composite resins *Eur. J. Esthet. Dent.* **8** 582–91
[58] Catelan A, Guedes A P A, Suzuki T Y U, Takahashi M K, de Souza E M, Briso A L F and dos Santos P H 2015 Fluorescence intensity of composite layering combined with surface sealant to staining solutions *J. Esthet. Restor. Dent.* **27** S33–40
[59] Da Silva T, de Oliveira H, Severino D, Balducci I, Huhtala M and Gonçalves S 2014 Direct spectrometry: a new alternative for measuring the fluorescence of composite resins and dental tissues *Oper. Dent.* **39** 407–15
[60] Giana H E, Silveira L, Zângaro R A and Pacheco M T T 2003 Rapid identification of bacterial species by fluorescence spectroscopy and classification through principal components analysis *J. Fluoresc.* **13** 489–93
[61] Walsh J D, Hyman J M, Borzhemskaya L, Bowen A, McKellar C, Ullery M, Mathias E, Ronsick C, Link J and Wilson M 2013 Rapid intrinsic fluorescence method for direct identification of pathogens in blood cultures *mBio* **4** e00865-13

[62] Dartnell L R, Roberts T A, Moore G, Ward J M and Muller J-P 2013 Fluorescence characterization of clinically-important bacteria *PLoS One* **8** e75270

[63] Lennon A, Buchalla W, Brune L, Zimmermann O, Gross U and Attin T 2006 The ability of selected oral microorganisms to emit red fluorescence *Caries Res.* **40** 2–5

[64] Volgenant C M, Hoogenkamp M A, Krom B P, Janus M M, Ten Cate J M, de Soet J J, Crielaard W and van der Veen M H 2016 Red and green fluorescence from oral biofilms *PLoS One* **11** e0168428

[65] Hope C, de Josselin de Jong E, Field M, Valappil S and Higham S 2011 Photobleaching of red fluorescence in oral biofilms *J. Periodont. Res.* **46** 228–34

[66] Hope C K, Billingsley K, de Josselin de Jong E and Higham S M 2016 A preliminary study of the effects of pH upon fluorescence in suspensions of prevotella intermedia *PLoS One* **11** e0158835

[67] Zhu H, Lao W, Chen Q, Zhang Q and Chen H 2015 A study of the fluorescence characteristics of common cariogenic microorganisms *Int. J. Clin. Exp. Med.* **8** 3651

[68] Kang S M, de Josselin de Jong E, Higham S M, Hope C K and Kim B I 2020 Fluorescence fingerprints of oral bacteria *J. Biophotonics* **13** e201900190

[69] Ammor M S 2007 Recent advances in the use of intrinsic fluorescence for bacterial identification and characterization *J. Fluoresc.* **17** 455–9

[70] Fyrestam J, Bjurshammar N, Paulsson E, Johannsen A and Östman C 2015 Determination of porphyrins in oral bacteria by liquid chromatography electrospray ionization tandem mass spectrometry *Anal. Bioanal. Chem.* **407** 7013–23

[71] Benson R, Meyer R, Zaruba M and McKhann G 1979 Cellular autofluorescence—is it due to flavins? *J. Histochem. Cytochem.* **27** 44–8

[72] Tani K, Watari F, Uo M and Morita M 2003 Discrimination between composite resin and teeth using fluorescence properties *Dent. Mater. J.* **22** 569–80

[73] Zhao B, Van Der Mei H C, Subbiahdoss G, de Vries J, Rustema-Abbing M, Kuijer R, Busscher H J and Ren Y 2014 Soft tissue integration versus early biofilm formation on different dental implant materials *Dent. Mater.* **30** 716–27

[74] Ko C-C, Yi D-H, Lee D J, Kwon J, Garcia-Godoy F and Kwon Y H 2017 Diagnosis and staging of caries using spectral factors derived from the blue laser-induced autofluorescence spectrum *J. Dent.* **67** 77–83

[75] Guo J Y, Zhang W L, Rao Y J, Zhang H H, Ma R, Lopes D S, Lins I C and Gomes A S 2020 High contrast dental imaging using a random fiber laser in backscattering configuration *OSA Contin.* **3** 759–66

[76] Tessaro L W, Dotta B T and Persinger M A 2019 Bacterial biophotons as non-local information carriers: species-specific spectral characteristics of a stress response *MicrobiologyOpen* **8** e00761

[77] Borisova E, Uzunov T and Avramov L 2006 Laser-induced autofluorescence study of caries model *in vitro Lasers Med. Sci.* **21** 34–41

[78] Borisova E G, Uzunov T T and Avramov L A 2004 Early differentiation between caries and tooth demineralization using laser-induced autofluorescence spectroscopy *Lasers Surg. Med.* **34** 249–53

[79] Brokos I, Stavridakis M, Lagouvardos P and Krejci I 2021 Fluorescence intensities of composite resins on photo images *Odontology* **109** 615–24

[80] Kurihara E, Koseki T, Gohara K, Nishihara T, Ansai T and Takehara T 2004 Detection of subgingival calculus and dentine caries by laser fluorescence *J. Periodont. Res.* **39** 59–65

[81] Karlsson L 2010 Caries detection methods based on changes in optical properties between healthy and carious tissue *Int. J. Dent.* **2010** 270729
[82] Tüfekçi E, Pennella D R, Mitchell J C, Best A M and Lindauer S J 2014 Efficacy of a fluoride-releasing orthodontic primer in reducing demineralization around brackets: an *in-vivo* study *Am. J. Orthod. Dentofac. Orthop.* **146** 207–14
[83] Fontana M, Platt J, Eckert G, González-Cabezas C, Yoder K, Zero D, Ando M, Soto-Rojas A and Peters M 2014 Monitoring of sound and carious surfaces under sealants over 44 months *J. Dent. Res.* **93** 1070–5
[84] Alammari M, Smith P, de Jong E d J and Higham S 2013 Quantitative light-induced fluorescence (QLF): a tool for early occlusal dental caries detection and supporting decision making *in vivo J. Dent.* **41** 127–32
[85] Markowitz K, Stenvall R and Graye M 2012 The effect of distance and tooth structure on laser fluorescence caries detection *Oper. Dent.* **37** 150–60
[86] Lennon A, Attin T and Buchalla W 2007 Quantity of remaining bacteria and cavity size after excavation with FACE, caries detector dye and conventional excavation *in vitro Oper. Dent.* **32** 236–41
[87] Ayad M F 2001 Effects of rotary instrumentation and different etchants on removal of smear layer on human dentin *J. Prosthet. Dent.* **85** 67–72
[88] Tung O-H, Lee S-Y, Lai Y-L and Chen H-F 2008 Detection of subgingival calculus through oral gum *in vitro* using two-photon fluorescence microscopy *30th Annual Int. Conf. of the Engineering in Medicine and Biology Society* (Piscataway, NJ: IEEE) pp 4051–4
[89] Laky M, Laky B, Arslan M, Lettner S, Müller M, Haririan H, Husejnagic S, Rausch-Fan X, Wimmer G and Moritz A 2021 Effectiveness of a 655-nm InGaAsP diode laser to detect subgingival calculus in patients with periodontal disease *J. Periodontol.* **92** 547–52
[90] Krause F, Braun A, Jepsen S and Frentzen M 2005 Detection of subgingival calculus with a novel LED-based optical probe *J. Periodontol.* **76** 1202–6
[91] Lussi A, Hibst R and Paulus R 2004 DIAGNOdent: An optical method for caries detection *J. Dent. Res.* **83** 80–3
[92] Castilho L S, Cotta F V, Bueno A C, Moreira A N, Ferreira E F and Magalhães C S 2016 Validation of DIAGNOdent laser fluorescence and the International. Caries Detection and Assessment System (ICDAS) in diagnosis of occlusal caries in permanent teeth: an *in vivo* study *Eur. J. Oral Sci.* **124** 188–94
[93] Shi X, Tranaeus S and Angmar-Mansson B 2001 Comparison of QLF and DIAGNOdent for quantification of smooth surface caries *Caries Res.* **35** 21
[94] Attrill D and Ashley P 2001 Occlusal caries detection in primary teeth: a comparison of DIAGNOdent with conventional methods *Br. Dent. J.* **190** 440–3

Chapter 4

ESR spectroscopy techniques for dentistry applications: recent advances

Rajeev Ranjan, Diplina Barman, Prashant Gupta and Sandeep Kumar

Electron spin resonance (ESR) imaging is a modern imaging technique with multiple medical and dental applications. ESR spectroscopy can be considered as an extended nuclear magnetic resonance (NMR) spectroscopy technique where the signaling subunits are unpaired electrons. ESR spectroscopy acts as a tool to identify free radicals (FRs) and study their structure. Therefore, ESR spectroscopy has proved to be a useful tool to select dental restoratives and assist in the design of improved restoratives. ESR also helps in detecting the mechanics behind the bone re-modeling that remains frozen in the existing bone matrix and is usually lost when new bone replaces old. ESR spectroscopy helps in detecting and quantifying the FRs of a sample. A microwave scanning ESR microscope studies the patterns of paramagnetic ions or radicals of materials present in the minerals. This chapter will review various aspects of ESR spectroscopy imaging and its application in dentistry.

4.1 Introduction

Electron spin resonance (ESR) spectroscopy can be considered as an extended nuclear magnetic resonance (NMR) spectroscopy technique where signaling sub-units are unpaired electrons. ESR spectroscopy is commonly used for scrutinizing organic radicals and metal complexes. ESR is also called electron paramagnetic resonance (EPR). The concept was initially discovered in Kazan State University by physicist Yevgeny Zavoiskyin 1944 and progressed at the University of Oxford, independently, by Brebis Bleaney. A lot of theories and mechanisms revolve around the concept of ESR or EPR. There are a couple of noted applications in the medical and dental fields. This chapter will review the various aspects of ESR spectroscopy imaging and its application in dentistry [1].

doi:10.1088/978-0-7503-3459-4ch4

4.2 ESR spectroscopy: theory

ESR spectroscopy is based on the magnetic properties of electrons and its association with intrinsic spin angular momentum. The electrons can be thought of as thin magnets. Organic molecules contain even electron numbers which make them uncharged. ESR spectroscopy analyzes molecules with odd electron numbers (unpaired electrons or free radicals (FRs)) [2]. The ESR method is suitable for examining systems with unpaired spins that interact with each other, stationary or time-dependent magnetic electric fields, coherent electromagnetic radiation and the magnetic moments of neighboring nuclei [3]. The results depend on the structure of the molecule, the distribution of spin densities, and the dynamics of molecules [3].

The set-up for an ESR spectrometer is shown in figure 4.1. Positioning FRs in the magnetic field initiates changes in the energy of the FRs based on the orientation of the electromagnetic moment and field direction. A two-directional orientation of the electron spin vector is permitted by quantum mechanics, namely the parallel and anti-parallel orientations. Resonance conditioning is the process of energy comprehension of the microwave radiation energy when FRs undergo transformation between two energy levels. The graphical representation of the absorption intensity at the energy level transition is known as the ESR spectrum (figure 4.2).

ESR spectroscopy acts as a tool to identify FRs and study their structure. Therefore, ESR spectroscopy has proved to be a useful tool to select dental restoratives and assist in the design of improved restoratives.

4.3 ESR spectroscopy in dentistry

ESR has broad applications in the fields of biology, medicine, the chemistry of radicals, metal organic compounds, solid state physics and chemistry, food chemistry, pharmacy, geology and archeology. In dentistry, ESR spectroscopy has

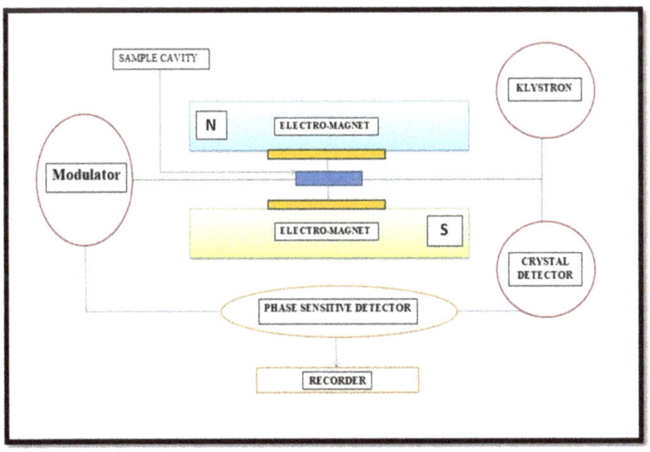

Figure 4.1. Simplified block diagram showing ESR spectroscopy.

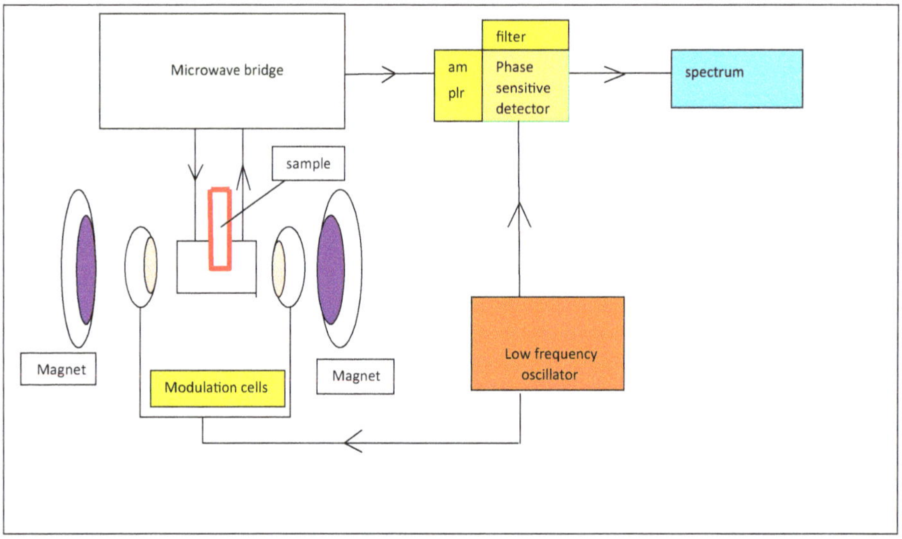

Figure 4.2. Simplified line diagram showing how signals are captured from samples exposed to magnetic fields and read as a graph or spectrum.

profound uses in the evaluation of the biology, kinetics, pathology, physiology and molecular kinetics of both hard and soft tissues. In the case of hard tissue, ESR spectroscopy has found applications in the evaluation of the curing kinetics of compomer type dental restoratives, determining of rate constants and half-lives of radicals. In the case of soft tissue, so far only limited research has been done in the field of detecting pathology and understanding the disease mechanisms in order to provide better treatment.

ESR imaging is a modern imaging technique with multiple medical and dental applications. The uses of ESR in dentistry are as follows:

- Bone imaging.
- Quantifying FRs in a sample.
- Determining the radical composition of human teeth.
- Determining the cumulative radiation dose of a tooth.
- Bone density detection.
- Evaluation of the curing kinetics of compomers.
- Studying the structure, morphology and sorption characteristics of samples of nanocrystalline C-HAP.
- Studying the light efficiency of a dental translucent fiberglass post to promote polymerization.
- Detecting the caries susceptibility of human tooth enamel.
- Detecting ROS in periodontitis of Down's syndrome
- Assessing salivary antioxidant levels.
- Studying the pathophysiology of TMDs.
- Studying cementum permeability.
- Monitoring nanoparticle endocytosis in cancer cells.

4.3.1 Bone imaging

ESR is useful in bone imaging due to the electron induced FRs that are responsible for source-specific stable ESR signals. ESR possesses the capability to trace temporarily presented integrated events. Reports have shown that, when used to study bone morphological patterns such as hydroxyapatite crystals, x-rays produce FRs. ESR imaging is useful for bone imaging because the electron induced FRs that are responsible for source-specific ESR signals are largely stable when trapped in bones and teeth. It can display temporarily integrated events from several years previous [12]. Investigators have the advantage of studying a person's past exposure to several source-specific types of radiation. The patterns vary according to the composition and configuration of the irradiated tissues. The selectively tuned effects of radiation exposure serve as diagnostic probes in iatrogenic events from known physiologic mechanisms. ESR also helps in detecting the mechanics behind bone re-modeling that remain frozen in the existing bone matrix and is usually lost when new bone replaces old [4].

ESR spectroscopy helps in detecting and quantifying the FRs in a sample. This is helpful for studying the degree of monomer-to-polymer conversion of dental composites. Electron spin spectroscopic analytical methods are also used to study the relationships between the structure, morphology and absorption characteristics of nanocrystalline carbonate-substituted hydroxyapatite (C-HAP) which is synthesized using an organic calcium source (poultry egg shells) acquired through chemical deposition of a solution [5].

4.3.2 Homogeneous catalysis research

EPR is successfully used in homogeneous catalysis research to characterize paramagnetic complexes and reactive intermediates. EPR investigates their electronic morphologies to understand their reactivity. Microscopic images of the radical concentration of human teeth irradiated by x-rays or gamma rays can also be obtained using an ESR microwave focused scanning microscope. A microwave scanning ESR microscope studies the patterns of paramagnetic ions or radicals of materials present in the minerals [5].

4.3.3 Cumulative radiation dose

The cumulative radiation dose of an *in vivo* tooth was measured without extraction using ESR dosimetry. The components of the system were:

i. A TE_{102} cavity with an aperture at the end plate.
ii. A Nd–B–Fe (Neomax) oral magnet.
iii. A commercial ESR spectrometer.

The front molar tooth was attached to the cavity to record the same [6].

An ESR probe includes a static field source and a microwave resonator for the measurement of paramagnetic defects in tooth enamel [7].

4.3.4 Bone density detection

Advancements in imaging techniques led to the development of newer techniques to measure radiation contamination or dose rates for individual doses. EPR measurements are based on the composition of hydroxyapatite mineral. A proportional relationship is observed between the radicals and the absorbed doses. The intensity of the resulting EPR absorption is the measure used for dose reconstruction. There is a difference in density for enamel and dentin. Preparation of the enamel samples plays an important role in EPR dose reconstruction. The main objective of preparation is to attain a minimal background signal with low inter-sample variation [1].

4.3.5 Evaluation of the curing kinetics of compomers

ESR spectroscopy can be used to study the mechanisms and kinetics of various polymer forming systems. These include photochemical polymerization and cross-linked acrylates or methacrylates. It can also be used to study polymer blends. EPR has been used in the magnetic resonance imaging of biological specimens, and also with nitroxide spin probes and in the characterization of irradiated and heated dental enamel. ESR has been used to obtain a variety of information including:

i. The progression of the conversion reaction due to light exposure.
ii. The time required for the decay of each radical type.
iii. The temperature variations and shattering effects on the materials under study.

ESR was seen to be helpful in monitoring and estimating the half-lives of propagating radicals in visible light initiated polymerizations. The ESR spectrum is used to measure the absorption of microwave radiation energy during the resonance condition (transition of unpaired electrons between two different energy levels) [5, 8].

4.3.6 The relationship between the structure, morphology and sorption characteristics of samples of nano-crystalline C-HAP

ESR spectroscopy applications in dosimetry have grown in importance recently. This method measures the radical concentration produced by ionizing radiation in the structure of the materials. Hydroxyapatite crystals consist of a majority of calcified tissues and hence serves as an important material for retrospective and accident dosimetry, the identification of irradiated meat containing bones and archeological dating. The marker systems are long-lived CO_2 radicals produced from CO_3 impurities that substitute PO_4 groups (B-type carbonated apatite) and/or OH groups (A-type carbonated apatite) in the hydroxyapatite lattice. Numerous studies have reported the nature of the species induced by radiation in the ESR spectra of synthetic apatites, but many fewer studies have reported the dosimetric potential of synthetic apatites [4, 9].

4.3.7 Detecting the radical concentration in human teeth

To detect the radical concentration in human teeth, the ESR probe includes a static field source and a microwave resonator for measuring the paramagnetic defects of tooth enamel. These defects serve as markers for quantifying the amount of ionizing radiation dose absorbed in the tooth. The probe measures the tooth when the probe is placed above its outer surface. It is operated at an ~6.2 GHz frequency in pulsed mode [10].

4.3.8 Studying the light efficiency of a dental translucent fiberglass post to promote polymerization

ESR spectroscopy facilitates the detection and quantification of FRs in a sample, and is associated with the degree of monomer-to-polymer conversion of dental composites.

A photo-polymerizable dental composite resin was maneuvered to trap all FRs detected by ESR spectroscopy. It was generated by light curing. The experiment involved the introduction of a 4 mm post which reduced the direct current (DC) by half compared to irradiation in the absence of the post. Additionally, increasing the length of the post drastically reduced the amount of FRs detected. This suggests that the translucent fiberglass may have absorbed a large portion of the radiation, and the DC of the resinous materials in the apical part of the root canal may not be sufficient to guarantee the fixation of the post [11].

4.3.9 Caries susceptibility in human tooth enamel

ESR helps in the detection of carbonate-related paramagnetic centers in human tooth enamel. The ESR spectrum signals of human bone were found to be better as compared to the other techniques determining caries susceptibility. Drastic changes in the spectra were noted when the samples were heated [8, 9].

4.3.10 Detecting reactive oxygen species (ROS)

ESR is very useful for detecting the ROS that accompany any kind of inflammation. ROS are generated by polymorphonuclear leucocytes that increase the oxidative stress and this remains a critical step during any kind of inflammation. ROS have been implicated as a cause for diseases such as periodontitis, temporomandibular disorders (TMDs) and impaired wound healing. Studies have reported ROS formation in Down's syndrome gingival fibroblasts that lead to periodontitis. These ROS were expressed more in Down's syndrome gingival fibroblasts than non-syndromic gingival fibroblasts. These studies have led to an understanding of the pathophysiological role of ROS in Down's syndrome that is promising for developing new treatment strategies such as including anti-oxidants that could address the ROS and these cases could thus be better managed [12].

4.3.11 Assessing salivary anti-oxidant levels

ESR spectroscopy can be used to assess the relationship between periodontal diseases, salivary antioxidant activity and various ROS. The antioxidant activity and cytokine levels in human body fluids are strongly associated with periodontitis. Saliva acts as a scavenger for superoxide and hydroxyl radicals (HO*). Research has reported reduced salivary superoxide and HO* scavenging activity and reduced interleukins in saliva in patients with periodontitis [13, 14].

4.3.12 Temporomandibular disorders (TMDs)

TMDs are commonly encountered and are difficult to assess and treat since they are multifactorial. Mechanical stress leads to oxidative stress of articular tissues in TMDs. The endothelial cells and synovial cells that are pertinent to arthritis cause oxidative damage when stimulated. These FRs in the pathophysiology of TMDs can be studied using ESR spectroscopy, which leads to better understanding of the disease pathophysiology and better clinical management. ESR studies have shown increased HO* in the presence of superoxide anion radicals. HO* detected in synovial fluid is generated by the biological Fenton reaction in which a superoxide anion radical (O_2*) can lead to the formation of H_2O_2 through a dismutation reaction by superoxide dismutase. The HO* can be generated from the reaction of resultant H_2O_2 with free iron ions [15].

4.3.13 Studying cementum permeability

Cementum permeability changes during the course of periodontal disease. ESR/EPR spectroscopy can be used to assess the transport of water soluble spin labeled molecules through cementum. Research by ESR spectroscopy has found large diffusion flux across dental hard tissues that have been exposed to a periodontal pocket environment, since the structure of dental hard tissue changes during the course of periodontal disease leading to a change in cemental permeability. This research facilitates planning new treatment options for periodontal diseases [16].

4.3.14 Monitoring of nanoparticle endocytosis in cancer cells

Nanoparticles are the object of research for uses in diagnostic and medical therapy. Iron-oxide nanoparticles are potential materials for use in medical imaging. The application of these nanoparticles as drug carriers is possible because of their ability to penetrate cell membranes. These particles can be characterized using ESR spectroscopy. This could be applied in the treatment of oral cancer as well [17] (figure 4.3).

4.4 Conclusion

Multiple application have been reported for the use of ESR spectroscopy in dentistry. There are numerous advantages that have been observed which increase the scope of research on these newer aspects of imaging techniques. The EPR dose responses were measured and remarkable responses were observed, thereby

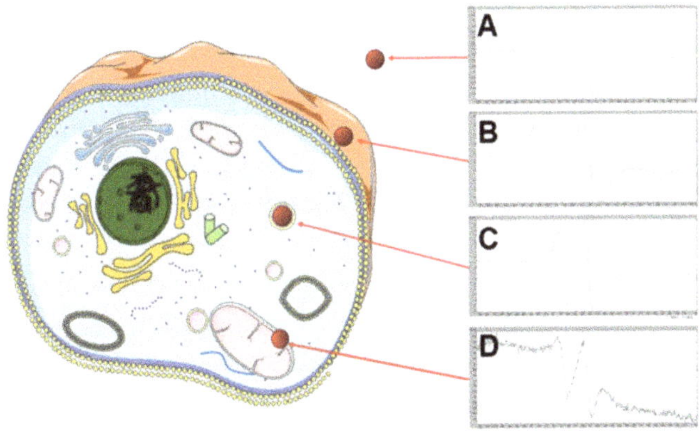

Figure 4.3. Changes in the structure and intensity of ESR spectra during the endocytosis process of functionalized nanoparticles in a cell. (A) ESR signal from the spin label not bonded to cells. (B) ESR signal from the spin label attached to a cell. (C) ESR signal from the spin label located inside cells (in organelles such as endosomes or lysosomes). (D) ESR signal from the spin label probably present in cellular mitochondria. Reprinted with permission from [17]. Creative Commons—Attribution 4.0 International—CC BY 4.0.

promoting ESR spectroscopy as one of the imaging techniques that have taken dentistry to a new level.

References

[1] Webber R L 1993 Oral imaging as a diagnostic tool for assessing osseous changes *J. Bone Miner. Res.* **8** S543–8

[2] Vickerman J C and Gilmore I S (ed) 2009 *Surface Analysis—The Principal Techniques* 2nd edn (New York: Wiley)

[3] Stößer R and Herrmann W 2013 Isotope effects in ESR spectroscopy *Molecules* **18** 6679–722

[4] Goloshchapov D L, Lenshin A S, Savchenko D V and Seredin P V 2019 Importance of defect nanocrystalline calcium hydroxyapatite characteristics for developing the dental biomimetic composites *Results Phys.* **13** 102158

[5] Adamczyk J, Ramos P and Pilawa B 2015 Effect of microwave power on EPR spectra of natural and synthetic dental biocompatible materials *Nukleonika* **60** 449–53

[6] Wieser A, El-Faramawy N and Meckbach R 2001 Dependencies of the radiation sensitivity of human tooth enamel in EPR dosimetry *Appl. Radiat. Isot.* **54** 793–9

[7] Umakoshi M *et al* 2017 *In vivo* electron paramagnetic resonance tooth dosimetry: dependence of radiation-induced signal amplitude on the enamel thickness and surface area of *ex vivo* human teeth *Health Phys.* **113** 262–70

[8] Tiba A, Culbertson B M and Berliner L J 1998 Electron spin resonance (ESR) spectroscopy studies of compomer type dental restoratives *J. Macromol. Sci.—Pure Appl. Chem.* **35** 1445–57

[9] De Oliveira L M, Rossi A M, Lopes R T, Rodrigues L N, Horowitz Y S and Oster L 2002 The influence of unstable signals for electron spin resonance dosimetry with synthetic A-type carbonated apatite *Radiat. Prot. Dosimetry* **101** 539–44

[10] Hochi A, Furusawa M and Ikeya M 1993 Applications of microwave scanning ESR microscope: human tooth with metal *Appl. Radiat. Isot.* **44** 401–5
[11] Vicentin B L S *et al* 2019 Application of ESR spectroscopy in the study of the light efficiency of a dental translucent fiberglass post to promote polymerization *J. Appl. Spectrosc.* **86** 244–9
[12] Komatsu T, Lee M C, Miyagi A, Shoji H, Yoshino F, Maehata Y, Maetani T, Kawamura Y, Ikeda M and Kubota E 2006 Reactive oxygen species generation in gingival fibroblasts of Down syndrome patients detected by electron spin resonance spectroscopy *Redox Rep.* **11** 71–7
[13] Yoshino F *et al* 2012 Assessments of salivary antioxidant activity using electron spin resonance spectroscopy *Arch. Oral. Biol.* **57** 654–62
[14] Tamaki N *et al* 2015 Relationship among salivary antioxidant activity, cytokines, and periodontitis: the Nagasaki Island study *J. Clin. Periodontol.* **42** 711–8
[15] Kawai Y, Kubota E and Okabe E 2000 Reactive oxygen species participation in experimentally induced arthritis of the temporomandibular joint in rats *J. Dent. Res.* **79** 1489–95
[16] Petelin M, Skaleric U, Cevc P and Schara M 1999 The permeability of human cementum *in vitro* measured by electron paramagnetic resonance *Arch. Oral. Biol.* **44** 259–67
[17] Krzyminiewski R, Dobosz B, Krist B, Schroeder G, Kurczewska J and Bluyssen H A R 2020 ESR method in monitoring of nanoparticle endocytosis in cancer cells *Int. J. Mol. Sci.* **21** 4388

IOP Publishing

Spectroscopic Techniques for Dentistry Applications
Recent advances
Ashutosh Kumar Shukla

Chapter 5

EPR spectroscopy in stomatology—towards and beyond dosimetry

Agata M Niewczas and Przemysław M Płonka

Tooth-based dosimetry of ionizing irradiation belongs to the most traditional applications of electron paramagnetic resonance (EPR) in biomedical research, originating directly from crystallography-oriented EPR. However, there are emerging new fields of EPR in stomatology focused on the free radicals generated by the living tissue itself, under normal conditions and in pathologies, including cancer and periodontitis. A separate, quickly developing EPR field concerns biomaterials and composites, including their polymerization and properties. This chapter gives an overview of these fields against the background of the biological and medical characteristics of material, taken from the most important handbooks in the field. This aims to make possible the ample use of EPR technology in stomatology both by dentists and by biophysicists who may not be aware of the specifics of the field.

5.1 Introduction

Shortly after its discovery in 1940s, the technique of electron paramagnetic resonance spectroscopy (EPR, also known as electron spin resonance (ESR)) (Kon 1989) was successfully applied to investigate crystals and their defects (Plonka 2019). The term 'paramagnetic center' (Pulatowa 1963a, 1963b) was initially used to describe one of the many types of crystal centers, determining their unique properties, mainly as crystal defects, appearing through interactions of a crystal with ionizing radiation (Plonka 2019, Atkins and Symons 1967). Consequently, EPR spectroscopy was applied to investigate crystal defects in teeth and has become a trusted technique for determining ionizing radiation dosimetry. EPR quickly entered the biological and medical sciences. EPR spectroscopists looked initially for systems and materials like the ones already investigated, among them finding the components of bones and teeth. The story of the application of EPR in dentistry is as long as its application in biology and medicine (Harshman and Johnson 2018).

doi:10.1088/978-0-7503-3459-4ch5

The human oral cavity does not consist solely of teeth, and stomatologic pathology does not concern only diseases of the teeth. Due to the solid-state nature of teeth, EPR measurement of this material is relatively easy (Rana *et al* 2010). We would like, however, to also focus on the diseases of the living tissues of the oral cavity, which can also undergo EPR measurements, in particular as the more recent literature devoted to teeth investigations is substantial (Satpathy *et al* 2019, Vicentin and Di Mauro 2019).

5.2 Principles of EPR

EPR is a type of magnetic resonance related to nuclear magnetic resonance (NMR). The phenomenon results from the magnetic properties of elementary particles, which for atoms, molecules and ions are responsible for their external magnetic properties. This is tightly coupled with spin, a property of the elementary particles—electrons, protons and neutrons (Gerlach and Stern 1922, Pauli 1925, Uhlenbeck and Goudsmit 1926). The EPR phenomenon is a result of the appearance of 'unpaired' electrons—the electrons of uncompensated electron spins—and sometimes also of the magnetic properties of the nucleus and its spin–orbital coupling. NMR and EPR are often confused by non-specialists (Płonka 2009), but they represent different interactions, differing in energy by 2–3 orders of magnitude.

The species responsible for the EPR phenomenon are all the atoms, ions and particles possessing an odd number of electrons, or an even number of electrons but with uncompensated spins (occupying different orbitals) (Płonka 2019). For such species the total energy is degenerated, and this degeneracy disappears when placing such a species into a stable external magnetic field. For such properties additional energy levels appear (the so-called Zeeman levels (Zeeman 1897)) and the substance, because of the transitions between these levels, can additionally absorb energy (usually microwave energy). This may result in the presence of additional spectral lines in their spectra of emission or absorption. This additional absorption of microwaves is specific to a given type of paramagnetic center and can be measured. This intensification of energy absorption when in an external magnetic field and after fulfilling some selection rules is called electron paramagnetic (spin) resonance (Zeeman 1897, Wertz and Bolton 1986, Plonka 2009).

The EPR line—the line of absorption—delivers information on the number and type of paramagnetic centers. Additionally, due to magnetic interaction with nuclei and the environment, it can deliver information on relaxation phenomena (the efficiency in dispersing the absorbed resonance energy) and on the properties of the local environment, which is valuable knowledge (Wertz and Bolton 1986, Plonka 2009). Finally, by special application of external field gradients, one can also obtain information on the distribution of the paramagnetic centers in the sample and the spatial distribution of their interactions with the microenvironment by means of EPR imaging (EPRI) (Elas *et al* 2019).

Due to the importance of the reactions and other phenomena related to unpaired electrons in biological systems, the EPR methodology has become an important method for the investigation of life in general (Płonka 2019).

5.3 Specificity of stomatology-derived materials for EPR measurements

5.3.1 A tooth and its components

The oscillations of the molecules of waters are energetically within the specific energies of the Zeeman transitions, which makes this phenomenon very sensitive to the liquid water contained in the sample. This lowers the sensitivity of EPR-based methods when applied to strongly hydrated samples (e.g. living systems) and limits the applications of EPR in biology (Rinard *et al* 2002). However, such applications are not only possible, but are also being actively developed and progressed. Nevertheless, samples with a low liquid water content are particularly suitable for EPR measurement. The mineralized bones and teeth are such materials, but they also contain some water and differ in their content of water. Table 5.1 contains the average water content in various parts of a tooth.

Teeth are the most frequently EPR-investigated stomatological materials. A tooth consists of four different layers (parts) and all of them (enamel, dentin, pulp and cementum) are a subject of EPR measurements. The literature devoted to tooth-based ionizing radiation dosimetry is very substantial and growing continuously (e.g. Vicentin and Di Mauro 2019, Satpathy *et al* 2019). Here, we characterize the parts of the tooth and give selected examples of their application as materials for EPR analysis.

5.3.1.1 Enamel

Tooth enamel covers the tooth's crown and determines its shape. It is the most mineralized tissue in the human body. It protects the dentin and the pulp of the tooth. Depending on the location within the crown, it has a thickness of 2–2.5 mm (Kmieć 2007). Enamel is formed in the process of amelogenesis as the only hard tissue of the tooth of epithelial origin. Ameloblasts—enamel-forming cells—are formed from the inner epithelium of the enamel-forming apparatus under the induction of the first predentin layer (Schröder 2000). The basic structural units of enamel are rods built of hydroxyapatites. There are 20 000–30 000 rods per 1 mm^2 (Kmieć 2007). The boundaries between the rods are visible due to the presence of interrod enamel. It is just as hard as the rods, but of a slightly different structure.

Table 5.1. Hydration of the main layers of a tooth; n.d. = no data. (After Kmieć (2007)).

	% mass	% volume
Enamel	3	9
Dentin	10	21
Pulp	75	n.d.
Cementum	10	n.d.

Interrod-building hydroxyapatites are arranged at an angle to the hydroxyapatites in the rods, which changes their optical properties.

According to Avery (2002) inorganic compounds constitute 96% by weight and 89% by volume of mature enamel. The main ingredient is calcium phosphate in the form of large crystals of dihydroxyapatite ($Ca_{10}(PO_4)_6(OH)_2$) and small amounts of calcium carbonate, calcium fluoride and magnesium phosphate. The enamel is the hardest on the surface due to the higher concentration of calcium and phosphates as well as fluorine and chlorine. Fluorine anions can replace hydroxyl ions. In this way, fluoroapatites are formed, which are more resistant to cariogenic agents. In the literature, we also find the concept of the mineral density of enamel, i.e. the content of hydroxyapatites in its volume. This value is variable and decreases from the surface towards the enamel–dentin border (Avery 2002).

Organic components constitute up to 1% of the weight of the enamel and up to 2% of the volume. They are mainly enamelin proteins. They are placed in the interprismatic space and fit closely to the prisms (Avery 2002).

The third component of enamel is water. It makes up 3% by weight and 9% by volume.

5.3.1.2 Dentin

Unlike enamel, dentin is a living tissue composed mainly of small, thin flakes of apatite crystals embedded in a protein matrix of cross-linked collagen fibers. It is slightly harder than bone, but does not show the ability to constantly remodel its structure. It is a support for the brittle enamel, and its elastic nature helps to neutralize stresses appearing in the tooth, e.g. during the act of chewing (Summit *et al* 2006). Both the pulp of the tooth and the dentin are formed from the mesenchymal tissue that forms the papilla dentis. Dentin develops from dentin-forming cells, odontoblasts, in the process of dentinogenesis. Dentin has a tubular structure. The S-shaped dentinal tubules run from the tooth pulp towards the enamel. Unlike enamel, the dentin substrate varies greatly with location, age and response to external stimuli. Due to the period of dentin production during an individual life, one can distinguish the primary dentin produced during tooth growth, the secondary physiological dentin, which is formed after the completion of tooth development, and the secondary pathological repair dentin, formed as a reaction of the pulp to pathological factors (Kmieć 2007, Nanci 2003). Mature dentin consists of 70% by weight and 47% by volume inorganic compounds. It is mainly calcium phosphate and trace amounts of fluorine and carbonate ions. The organic compounds are mainly collagen type I, phosphoproteins, proteoglycans and phospholipids, alkaline phosphatase, metalloproteinases and blood serum proteins. Organic compounds constitute 20% by weight and 32% by volume of the dentin. Water makes up 10% by weight and 21% by volume of this part of teeth (Avery 2002).

5.3.1.3 Pulp

Dentin is an integral part of the pulp–dentin complex (endodontium). The endodontium is a structural and functional unit of the tooth (Hargreaves *et al* 2011, Ingle *et al* 2002). The dental pulp fills the tooth cavity and the root canal. The tissue ends in the physiological foramen near the root apex. The pulp of the tooth is

75% water and 25% organic components (Kmieć 2007). The structure of the pulp resembles the embryonic connective tissue, including arterioles, venules, myelinated and unmyelinated nerves, and lymphatic channels bunded into trunks. It has a layered structure. There are three layers: the most peripheral layer is the odontoblastic layer. Below this is the subodontoblastic layer, a cell-free zone (the Weil layer), and the third layer is a cell-rich zone. As long as the tooth remains viable, they produce and adapt the dentin matrix to include collagen fibers and take part in the active transport of calcium ions. The cell-free zone consists mainly of numerous cytoplasmic extensions of the fibroblasts of the cell-rich zone and collagen fibers lying mainly along the nerve fibers. We also find single fibroblasts there. The cell-rich zone mainly consists of fibroblasts, undifferentiated mesenchymal cells, stem cells, pericytes, and smooth muscles cells supporting the capillary, and are all prime candidates as progenitor cells capable of differentiation and phenotype conversion into matrix-secreting replacements for destroyed primary odontoblasts. In the pulp one can also find immunocompetent cells such as macrophages, lymphocytes and dendritic cells. The pulp of the tooth has several functions: formative (the process of dentinogenesis), nutritive, sensory, and protective (coordinating responses to injury) (Kmieć 2007, Schröder 2000).

5.3.1.4 Cementum

Cementum is the calcified connective tissue that covers the root of the tooth. The thickness of the layer varies depending on the location. In the area of the tooth neck it is from 20 to 50 μm and at the root apex it increases to 150–200 μm. The organic components of cementum are cementoblasts, which form the organic matrix, and cementocytes. The inorganic components are mainly hydroxyapatite crystals, calcium carbonate and magnesium phosphate. About 10% of the weight of cementum is water.

Histologically there are four types of cementum:

- Acellular afibrillar cementum (AAC), located within the cemento–enamel junction.
- Acellular extrinsic fiber cementum (AEFC), located in the apical 1/3 of the root of the tooth.
- Acellular intrinsic fiber cementum (CIFC), located in the apical 2/3 of the root of the tooth.
- Cellular mixed stratified cementum (CMSC), located within the root apex and the furcations of multi-rooted teeth.

The basic functions of cementum are stabilization (keeping the tooth in the tooth-socket), repairing (repairing damage on the root surface) and protection (a barrier against harmful factors) (Kmieć 2007, Schröder 2000, Nanci 2003).

5.3.2 The applicability of the various layers of teeth for EPR measurement and dosimetry

Enamel, as the hardest, most crystallized and least hydrated material, is the best natural dosimeter, also because of its placement in the oral cavity and most periferally in the tooth, making it the best representative for the actual dose absorbed by a person (Doi 1986). The application of tooth enamel as a dosimeter for ionizing radiation is the most elementary example of the application of EPR to the investigation of biological materials (Harshman and Johnson 2018). Its low water content even enables using Q-band spectroscopy (Romanyukha *et al* 2014) and dosimetry *in vivo* or *ex vivo* using L-band and surface coils is also applicable (Zdravkova *et al* 2002, Umakoshi *et al* 2017). Tooth enamel may also be used in measurements requiring more sensitive detection, e.g. paramagnetic centers generated mechanically (Aragno *et al* 2001), thermally (Fattibene *et al* 2000) or by UV-irradiation (Liidja *et al* 1996), There are also more sophisticated examples of EPR applications, e.g. to determine the biological variability (thickness) of enamel (Martens *et al* 1986, Umakoshi *et al* 2017), the area of irradiation (Umakoshi *et al* 2017), or even the penetration of nitroxides (and consequently, of other free radicals) through enamel (Cevc *et al* 1986).

Dentin is also a good material for EPR analysis (Zdravkova *et al* 2002). The applications of EPR follow the examples of dosimetric analysis performed on enamel (Haskell *et al* 1995, Harshman *et al* 2018) but there are also quite numerous structural studies in which normal dentin is compared to deproteinated material (Fattibene *et al* 2006a, 2006b) or thermally generated EPR signals of dentin and enamel (Bachmann *et al* 2004). The techniques include simulations of EPR signals (Strancar and Petelin 2010).

In addition to classical dosimetry (Hayes *et al* 1998, Klevezal *et al* 2001), EPR is a useful tool for structural investigations of cementum (Scaleric *et al* 1998), which, as it is usually embedded in the soft tissues of the gums, differs in its dosimetric capacities. EPR is also a good tool to investigate the permeability of cementum (Petelin *et al* 1999) and to control the effects of laser irradiation (Gaspirc and Scaleric 2001).

5.4 Saliva

Despite limiting the efficiency of EPR measurements of biological systems, many works have been performed in a water environment, including chemical systems, *in vitro* cellular cultures, and normally hydrated living tissues *ex vivo* and *in vivo* (Plonka 2009). The most characteristic body fluid associated with the oral cavity and stomatology is saliva. It is secreted by the salivary glands. Based on their size, they are divided into the large and small salivary glands. In addition to the large parotid, submandibular and sublingual glands, there are small glands, whose number in the oral cavity reaches several hundred. The small salivary glands are located in the whole oral cavity apart from the gums and the front part of the hard palate. Based on the type of saliva secreted, the salivary glands are divided into purely serous (the parotid gland), purely mucous (numerous small glands in the mucosa) and mixed (the submandibular and sublingual glands, and some of the small salivary glands) (Kmieć 2007, Ferguson 1999).

Whole mixed saliva is considered to be the natural environment of the oral cavity. It is a mixture of the secretions from all the salivary glands. During the day, the amount of saliva produced varies and ranges from 0.5–1 l. In an adult, about 80% is stimulated saliva secreted during the act of chewing or speaking. The remaining 2%–10% is produced at night (unstimulated) (Ferguson 1999).

The main component of saliva is water (as much as 99%). The remaining 1% is organic solids, mainly proteins, glycoproteins and various enzymes, and inorganic components, such as calcium, phosphate, carbonate, sodium, potassium, chloride and fluoride ions (Ferguson 1999).

Saliva is a key element in maintaining the integrity of the hard and soft tissues and the health of the oral cavity. It has many functions. They depend on both its properties and solid components. The main functions of saliva are: protective, related to food intake and related to speech (Fejerskov *et al* 2015).

5.4.1 EPR measurements of saliva

All the substances which appear in the oral cavity, and also all the products of biochemical reactions that take place there, can be found in saliva. If they are paramagnetic, it gives the opportunity to detect them directly with EPR (Goodman *et al* 2002, Yoshino *et al* 2012). They may be also trapped by spin traps, which makes it possible to increase their stationary level over the limit of EPR detectability (Plonka 2009). A good and important example is EPR spin trapping of nitric oxide (NO) in saliva (Takahama *et al* 2008). NO produced during inflammation mainly by cytotoxic activated macrophages is one of the important factors of periodontosis (Wang *et al* 2019). It creates stable complexes with hemoglobin (Dutka *et al* 2019) which are also detectable by EPR, and affects hemoglobin itself, e.g. participating in its oxidation and redox properties (Rifkind *et al* 2011). Perhaps it is one of the factors that degrades or modifies heme proteins in periodontosis (Sroka *et al* 2001). It turns out that NO generated by nitrite-reducing oral microbiome in the oral cavity can affect blood pressure (Bryan *et al* 2017), thus acting systemically on the whole organism. EPR is an important research technique to measure NO in saliva.

5.5 Mucosus membrane

The oral cavity consists of the vestibule (*vestibulum oris*) and the oral cavity (*cavum oris proprium*). The mucosa lines the entire oral cavity, however, it varies significantly in terms of structure and function depending on its location. There are three types of mucosa in the oral cavity (Kmieć 2007):

1. Chewing mucosa covers an area of approximately 25%, covering fixed structures such as the gums and hard palate, and is resistant to the compressive forces and friction occurring during the act of chewing.
2. The lining mucosa accounts for approximately 60% of the total surface of the oral cavity, and is located within the lips, cheeks, the floor of the mouth and the soft palate. It is movable and has a simplified structure.
3. The special mucosa makes up only about 15% of the surface of the oral cavity, covering the tongue ridge, and is responsible for perceiving pain, taste and temperature stimuli (Kmieć 2007).

Histologically, the entire oral cavity is lined with a multilayered planar epithelium. The basic structural units of this epithelium are keratinocytes. These cells are very resistant to mechanical and chemical factors. During development, keratinocytes differentiate into keratinized and non-keratinized epithelium.

The multilayered planar keratinizing epithelium covers the chewing mucosa. It consists of four layers: *stratum basale*, *stratum spinosum*, *stratum granulosum* and *stratum corneum*.

Multilayered planar non-keratinized epithelium covers the lips, cheeks, floor of the mouth, soft palate and lower surface of the tongue. Like the keratinizing epithelium, it consists of four layers: *stratum basale*, *stratum spinosum*, *intermediate layer* and *stratum disjunctum*.

There are also non-epithelial cells in the epithelium lining the mouth. These are melanocytes, Largenhans cells, Merkel cells and lymphocytes. They constitute about 10% of all the epithelial cells (Schröder 2000).

The main layer under the epithelial part of the oral mucosa is the *lamina propria mucosae*. It is a flabby connective tissue rich in cells and collagen fibers. It also contains blood and lymph vessels, nerves and small salivary glands.

There is a submucosa in the parts of the mouth that are covered with the lining mucosa. It has a much looser structure than the *lamina propria mucosae*. It contains larger arteries, veins and also small salivary glands.

The main functions of the mucosa are protective, renewing, sensory and secretory (Schröder 2000, Wolf *et al* 2005).

5.5.1 Oral mucosa in EPR

The cells of the mucosal membrane may undergo neoplastic transformation and be the origin of cancer. The factors which can lead to such transformation include EPR-detectable free radicals produced during inflammation (in addition to NO, mainly superoxide, and various metabolites of these radicals). Free radicals may appear in the oral cavity from food, and its mechanical or enzymatic transformation (Gallez *et al* 2000, Goodman *et al* 2002). When writing about cancer itself, an important factor determining the effectiveness of the therapy, mainly the radiotherapy, is the level of oxygen, which may be monitored or even imagined by EPR (Erjavec *et al* 2006, Bratasz *et al* 2007, Chen *et al* 2021). Another important application of EPR in oral therapy is monitoring of drug delivery in the form of mucoadhesive ointments or nanocarriers (Petelin *et al* 1998, 2004, Erjavec *et al* 2006, Jager *et al* 2018).

A rare but deadly neoplasm of the oral cavity is melanoma (Malinoski *et al* 2019), a very aggressive tumor (de Resende *et al* 2021) originating from infrequent melanocytes which may appear in the oral mucosa (de Lima Morais *et al* 2020, Feller *et al* 2014, Lin *et al* 2020) or maxillary gingivae (Shah *et al* 2021). Other oral tumors may, although rarely, present a pigmented phenotype as well (de Morais *et al* 2018, Soares *et al* 2021). Melanin is a paramagnetic biopolimer (Commoner *et al* 1954) appearing in the oral cavity, in addition to melanocytes, also in the epithelial cells (Fricaid and Sibaud 2017). It may strongly affect the resistance of a tumor to therapy (Slominski *et al* 2015). On the other hand, its stable free-radical-like EPR signal is a precious source of information concerning melanoma cells and tissue (Plonka 2009, Godechal *et al* 2013), which makes this technique an important tool in the fight against oral melanoma (Al Khatib *et al* 2020).

5.6 Biomaterials

Biomaterials possess wide clinical applications. They are used in the treatment of tooth decay as well as in endodontic treatment, internal resorption and chronic diseases of the periapical tissues (Ingle *et al* 2002).

The most commonly used biomaterials include preparations based on calcium hydroxide. They strongly alkalinize the environment (pH 12.5) through the release of hydroxyl ions, OH^-. As a result, they induce repair processes and calcification in the tooth pulp. They stimulate odontoblasts to produce a reparative dentin with an irregular structure—the dentin bridge. Due to its high pH, calcium hydroxide also has an antibacterial effect, neutralizing the lactic acid produced by osteoclasts and activates the enzyme alkaline phosphatase, which is involved in the formation of hard tissues (Ingle *et al* 2002).

Another example is mineral trioxide aggregate (MTA). It consists of three oxides: calcium, silicon and aluminum. Like calcium hydroxide, it has a high, alkaline pH (from 10.2 to 12.5). The mechanism of action of MTA is not fully understood. The therapeutic effect is achieved by changing the trioxide to calcium hydroxide. The dentin bridge produced by the action of MTA grows faster and is thicker, although it has a relatively low antibacterial effect (Hargreaves *et al* 2011).

Another bioactive preparation is biodentin. It is a calcium silicate cement. It is produced with the use of active biosilicates. It is made when a powder (tricalcium silicate, calcium carbonate, zirconium dioxide) is mixed with a liquid containing calcium chloride, a soluble polymer and water. It is a dentin substitute. Like calcium hydroxide and MTA, it has an antibacterial effect and stimulates odontoblasts to form reparative dentin (Ingle *et al* 2002).

5.6.1 Resin composites

Resin composites were first used in the 1950s and 1960s by the 'legendary' Bowen (1956, 1982). These were autocured materials based on the oligomeric bis-GMA (bisphenol A-glycidyl methacrylate), resin and bisphenol A. Silica, quartz and glass were used as fillers. Since then, composite materials have been constantly improved both in terms of composition (other types of resin were used, the size of the filler particles was significantly reduced) and the method of polymerization (light polymerization using halogen and LED lamps) (Anusavice *et al* 2013).

Currently, light-cured composites are widely used. They consist of four main components:

1. An organic resin matrix, based on a blend of aromatic and/or aliphatic dimethacrylate monomers such as bis-GMA and urethane dimethacrylate (UDMA).

2. An inorganic filler—amorphous silica, quartz, barium and strontium glass, fluorosilicates, ytterbium trifluoride and ytterbium fluoride—generally between 30% to 70% by volume or 50% to 85% by weight of a composite.
3. A coupling agent, γ-methacryloxypropyltrimethoxysilane is used most commonly.
4. An initiator/activator system.

There are many classifications of composite materials, but one of the most useful is the size division of the filler particles (Anusavice *et al* 2013, McCabe *et al* 1998).

Composite polymerization takes place in the initiator–activator process. It is an addition polymerization initiated by free radicals that are activated by visible blue light. The peak of the wave is from 470 to 480 nm. The light is absorbed by the photoinitiator camphorquinone. The reaction is accelerated by the organic amine. The degree of polymerization depends on the exposure time and the distance of the light source from the composite (Anusavice *et al* 2013, McCabe *et al* 1998).

The disadvantages of composite materials are polymerization shrinkage, sensitivity to daylight, limited depth of polymerization (the optimal layer is 2 mm) and the need to select an appropriate polymerization lamp. Many of the light-curing composites in their pure form are cytotoxic *in vitro*. Their release into the tissues depends on the type of composite and the degree of polymerization. Under clinical conditions, composites release their components for a few more weeks. It is believed that dentin is a natural barrier that significantly limits their penetration into the tooth pulp. However, if it breaks, these components have a negative effect on the pulp of the tooth. There is no clear evidence as to how considerable an influence the low concentrations of toxic composite components have in terms of a negative effect on the pulp. The components of the polymerized composite are also released into the oral cavity environment. However, there are no studies confirming the negative effects of their action (Anusavice *et al* 2013).

The advantages of light-cured composite materials include high esthetic value, the possibility of use in many clinical situations, the working time controlled by a dentist before final hardening, easy application, and polymerization in small batches, which significantly reduces the amount of polymerization shrinkage (Anusavice *et al* 2013, McCabe *et al* 1998).

5.6.2 EPR in the research of biomaterials and composites used in dentistry

As the mechanism of polymerization of biomaterials is often free-radical (Vicentin *et al* 2016) and/or photochemical (Lindén 1993, Vicentin *et al* 2019), thus as it generates free radicals anyway, EPR spectroscopy is an important method applied in this branch of stomatology, being actually in parallel a branch of technology (Kiteska *et al* 2018). EPR is used to monitor and analyze the process of polymerization on-line, and the interactions of the produced biomaterial with the alive environment in the long run. Some biomaterials are generated and have been designed under the strict control of EPR spectroscopy. An interesting example is the generation of polymers containing a biopolymer itself—melanin—which may not only become a new biomaterial but also a measure to investigate melanin properties in the solid state (Al Khatib *et al* 2020). This is a big domain of stomatology and biomedical EPR and numerous handbooks and chapters have been devoted to this topic (Vicentin and Di Mauro 2019, Satpathy *et al* 2019).

5.7 Conclusions—towards future stomatology

For the last 70 years EPR spectroscopy has spread its applicability over many disciplines of biology and medicine, gaining an independent but interdisciplinary character, similar to other '-omics' (Plonka 2019). Nevertheless, it still suffers methodological problems in *in vivo* applications and here one can expect the most important progress in stomatology-related EPR (Buchbinder *et al* 2021). In addition to its classical position in biomedical research, as a dosimetric technique it has gained a new meaning, in particular because of its paleontological and astro-biological uses. The 'fingerprints of life' are often of a paramagnetic character (Perrette *et al* 2015), and irradiation belongs to the most problematic aspects of the anticipated interplanetary travels (Vasileiou and Summerer 2020). In term of stomatological applications, beyond dosimetry, new fields of stomatological EPR have arisen, which may be divided into three main groups: (i) EPR in stomatological biomaterials and composites, being more of a branch of technology to follow the free-radical or light-mediated reactions of polymerization; (ii) EPR in stomatological oncology monitoring of free-radical factors which may cause cancer; and (iii) EPR in periodontosis—devoted to the mechanisms and paramagnetic products of soft tissue destruction in the oral cavity. In parallel to the development of EPR technology, new ideas for EPRI and oximetry are appearing in stomatology. The most interesting is the application of informatic technologies (artificial intelligence, deep learning algorithms, neural networks, etc) to modern stomatology. In recent papers on this topic (Nguyen *et al* 2021) one can find similar problems as in the EPR applications in stomatology, such as radiological periodontology, etc, which will inevitably lead to the fusion of stomatologic IT and EPR in the future.

Author contributions

AN contributed the stomatological part of the text (50%) and PMP the biophysical part of the text (50%); both authors participated equally in the preparation of the manuscript.

References

Al Khatib M, Costa J, Baratto M C, Basosi R and Pogni R 2020 Paramagnetism and relaxation dynamics in melanin biomaterials *J. Phys. Chem.* B **124** 2110–5

Anusavice K J, Phillips R W, Shen C and Rawls H R 2013 *Phillips' Science of Dental Materials* 6th edn (St Louis, MO: Elsevier/Saunders)

Aragno D, Fattibene P and Onori S 2001 Mechanically induced EPR signals in tooth enamel *Appl. Radiat. Isot.* **55** 375–82

Atkins P W and Symons M C R 1967 *The structure of inorganic radicals. An application of electron spin resonance to the study of molecular structure* (Amsterdam: Elsevier Publishing

Company); (Russian transl. 1970 Спектры ЭПР и строение неорганичецких радикалов ed Germanov E D, Dyatkina M E (Москва: Издателство 'Мир' p 56–65)
Avery J K 2002 *Oral Development and Histology* (Stuttgart: Thieme)
Bachmann L, Baffa O, Gomes A S L and Zezell D M 2004 Chemical origin of the native ESR signals in thermally treated enamel and dentin *Phys. B: Condens. Matter* **349** 119–23
Bowen R L 1956 Use of epoxy resins in restorative materials *J. Dent. Res.* **35** 360–9
Bowen R L 1982 Composite and sealant resins—past, present, and future *Pediatr. Dent* **4** 10–15
Bratasz A, Kulkarni A C and Kuppusamy P 2007 A highly sensitive biocompatible spin probe for imaging of oxygen concentration in tissues *Biophys. J.* **92** 2918–25
Bryan N S, Tribble G and Angelov N 2017 Oral microbiome and nitric oxide: the missing link in the management of blood pressure *Curr. Hypertens. Rep.* **19** 33
Buchbinder L, Datz H, Dayan N, Carmielli R and Blank A 2021 The potential of pulsed electron spin resonance for tooth-based retrospective biodosimetry *Appl. Magn. Reson.* https://doi.org/10.1007/s00723-021-01417-z
Cevc P, Zupancic I and Lavrencic B B 1986 The diffusion measurements of nitroxide spin labels in the tooth enamel by electron-paramagnetic-res imaging *Magnet. Reson. Med.* **3** 820–0
Chen E Y, Tse D, Hou H, Schreiber W A, Schaner P E, Kmiec M M, Hebert K A, Kuppusamy P, Swartz H M and Williams B B 2021 Evaluation of a refined implantable resonator for deep-tissue EPR oximetry in the clinic *Appl. Magn. Reson.* **52** 1321–42
Commoner B, Townsend J and Pake G 1954 Free radicals in biological materials *Nature* **174** 689–91
de Lima Morais T M D, Soares C D, Wanderley A E C, de Oliveira C R R, Ribeiro C M B, de Almeida O P, de Carvalho M G F, Ferreira S M S 2020 Oral melanomas in HIV-positive patients: Report of two cases and review of the literature *Oral Oncol.* **101** 104375
de Morais P M, Schettini A P M, Rocha J A and da Silva R C 2018 Pigmented squamous cell carcinoma: case report and importance of differential diagnosis *An. Bras. Dermatol.* **93** 96–08
de Resende T A C, de Andrade B A B, Bernardes V F, Coura B P, Delgado-Azanero W, Mosqueda-Taylor A, de Almeida O P, Gomes C C and Gomez R S 2021 BRAFV600E mutation in oral melanocytic nevus and oral mucosal melanoma *Oral Oncol* **114** 105053
Doi Y 1986 Electron-paramagnetic-res studies of tooth enamel *Magnet. Reson. Med* **3** 815
Dutka M, Pyka J and Płonka P M 2019 EPR studies on understanding the physical intricacy of HbNO complexes *Electron Spin Resonance in Medicine* ed A K Shukla (Singapore: Springer Nature) pp 23–43
Elas M, Krzykawska-Serda M, Gonet M, Kozińska A, Płonka PM 2019 Electron paramagnetic resonance imaging-solo and orchestra *Medical Imaging Methods* ed A K Shukla (Singapore: Springer) ch 1 pp 1–42
Erjavec V, Pavlica Z, Sentjurc M and Petelin M 2006 *In vivo* study of liposomes as drug carriers to oral mucosa using EPR oximetry *Int. J. Pharmaceut.* **307** 1–8
Fattibene P, Aragno D, Onori S and Pressello M C 2000 Thermal induced EPR signals in tooth enamel *Radiat. Meas* 32 793–8
Fattibene P, Carosi A, De Coste V and Onori S 2006a EPR properties of intact and deproteinated dentin *Radiat. Prot. Dosim.* **120** 216–20
Fattibene P, Carosi A, De Coste V, Sacchetti A, Nucara A, Postorino P and Dore P 2006b A comparative EPR, infrared and Raman study of natural and deproteinated tooth enamel and dentin *Phys. Med. Biol.* **50** 1095–108
Fejerskov O, Nyvad B and Kidd E 2015 *Dental Caries: The Disease and Its Management* 3rd edn (Oxford: Wiley Blackwell)

Feller L, Masilana A, Khammissa R A G, Altini M, Jadwat Y and Lemmer J 2014 Melanin: the biophysiology of oral melanocytes and physiological oral pigmentation *Head Face Med.* **10** 8
Ferguson D B 1999 *Oral Bioscience* (Edinburgh: Churchill Livingstone)
Fricain J C and Sibaud V 2017 Pigmentations of the oral cavity *Presse Med.* **46** 303–19
Gallez B, Baudelet C and Debuyst R 2000 Free radicals in licorice-flavored sweets can be detected noninvasively using low frequency electron paramagnetic resonance after oral administration to mice *J. Nutr* **130** 1831–3
Gaspirc B and Skaleric U 2001 Morphology, chemical structure and diffusion processes of root surface after Er:YAG and Nd:YAG laser irradiation *J. Clin. Periodontol.* **28** 508–16
Gerlach W and Stern O 1922 Das magnetische Moment des Silberatoms *Z. Phys.* **9** 353–5
Godechal Q, Ghanem E, Martin G C and Gallez B 2013 Electron paramagnetic resonance spectrometry and imaging in melanomas: comparison between pigmented and nonpigmented human malignant melanomas *Mol. Imaging* **12** 218–23
Goodman B A, Glidewell S M, Arbuckle C M, Bernardin S, Cook T R and Hillman J R 2002 An EPR study of free radical generation during maceration of uncooked vegetables *J. Sci. Food Agricult.* **82** 1208–15
Hargreaves K M, Cohen S and Berman L H 2011 *Cohen's Pathways of the Pulp* 10th edn (St. Louis, MO: Mosby)
Harshman A and Johnson T 2018 A brief review-EPR dosimetry and the use of animal teeth as dosimeters *Health Phys* **115** 600–7
Haskell E H, Kenner G H and Hayes R B 1995 Electron-paramagnetic-resonance dosimetry of dentin following removal of organic material *Health Phys* **68** 579–84
Hayes R B, Kenner G H and Haskell E H 1998 EPR dosimetry of Pacific walrus (*Odobenus rosmarusdivergens*) teeth *Radiat. Prot. Dosim.* **77** 55–63
Ingle J I and Bakland L K 2002 *Ingle's Endodontics* 5th edn (London: BC Decker/Hamilton)
Jager J *et al* 2018 Characterization of hyperbranched core–multishell nanocarriers as an innovative drug delivery system for the application at the oral mucosa *J. Periodont. Res.* **53** 57–65
Kiteska B, Funduk N, Cevc P, Jesih A, Anzlovar A and Kopac I 2018 The influence of free-radical concentration on the shear bond strength dental composites *Mater. Tehnol.* **52** 177–82
Klevezal G A, Bakhur A E, Sokolov A A, Serezhenkov V A and Krushinskaya N K 2001 Retrospective estimation of radiation load and its effect on some biological parameters of reindeer (*Rangifer tarandus*) from Wrangel Island *Russ. J. Ecol.* **32** 110–6
Kmieć Z 2007 *Histologia i Cytofizjologia Zęba i Jamy Ustnej* 1st edn (Wrocław: Elsevier/Urban) (in Polish)
Kon H 1989 Recommendations for EPR/ESR nomenclature and conventions for presenting experimental data in publications *Pure Appl. Chem.* **61** 2195–200
Liidja G, Past J, Puskar J and Lippmaa E 1996 Paramagnetic resonance in tooth enamel created by ultraviolet light *Appl. Radiat. Isot.* **47** 785–8
Lin C W, Lee Y P, Hwang M J and Chiang C P 2020 Oral mucosal melanoma: case report *J. Dent. Sci.* **15** 556–7
Lindén L-Å 1993 Applied photochemistry in dental science *Proc. Indian Acad. Sci. (Chem. Sci.)* **105** 405–19
Malinoski H, Reddy R, Cohen D M, Bhattacharyya I, Islam M N and Bowers T L 2019 Oral melanomas: a case series of a deadly neoplasm *J. Oral Maxillofacial Surg.* **77** 1832–6
Martens lC, Verbeeck R M H, Callens F J and Matthys P F A 1986 Biological variations in sound human tooth enamel studied by electron-paramagnetic-res *J. Dent. Res.* **65** 552

McCabe J F and Angus W G 1998 *Applied Dental Materials* 8th edn (Oxford: Blackwell)
Nanci A 2003 *Ten Cate's Oral Histology. Development, Structure and Function* 6th edn (St Louis, MO: Mosby)
Nguyen T T, Larrivée N, Lee A, Bilaniuk O and Durand R 2021 Use of artificial intelligence in dentistry: current clinical trends and research advances *J. Can. Dent. Assoc.* **87** 17
Pauli W 1925 Über den Zusammenhang des Abschlusses der Elektronengruppenim Atom mit der Komplexstruktur der Spektren *Z. Phys* **31** 765–83
Perrette Y, Poulenard J, Protiere M, Fanget B, Lombard C, Miege C, Quiers M, Nafferchoux E and Pepin-Donat B 2015 Determining soil sources by organic matter EPR fingerprints in two modern speleothems *Org. Geochem.* **88** 59–68
Petelin M, Pavlica Z, Bizimoska S and Sentjurc M 2004 *In vivo* study of different ointments for drug delivery into oral mucosa by EPR oximetry *Int. J. Pharmaceut* **220** 83–91
Petelin M, Sentjurc M, Stolic Z and Skaleric U 1998 EPR study of mucoadhesive ointments for delivery of liposomes into the oral mucosa *Int. J. Pharmaceut.* **173** 193–202
Petelin M, Skaleric U, Cevc P and Schara M 1999 The permeability of human cementum *in vitro* measured by electron paramagnetic resonance *Arch. Oral Biol.* **44** 259–67
Plonka P M 2009 Electron paramagnetic resonance as a unique tool for skin and hair research *Exp. Dermatol* **18** 472–84
Płonka P M 2019 EPR Paramagnetomics *Electron Spin Resonance in Medicine* ed A K Shukla (Singapore: Springer) ch 9 pp 189–221
Pulatova M K 1963a Localization of paramagnetic centre in gama irradiated protein *Biophysics-USSR* **8** 700–3
Pulatova M K 1963b [On the problem of the localization of the paramagnetic center of proteins irradiated with gamma rays] *Biofizika* **8** 632–4 [Article in Russian]
Rana S, Chawla R, Kumar R, Singh S, Zheleva A, Dimitrova Y, Gadjeva V, Arora R, Sultana S and Sharma R K 2010 Electron paramagnetic resonance spectroscopy in radiation research: current status and perspectives *J. Pharm. Bioallied Sci.* **2** 80–7
Rifkind J M, Nagababu E and Ramasamy S 2011 The quaternary hemoglobin conformation regulates the formation of the nitrite-induced bioactive intermediate and the dissociation of nitricoxide from this intermediate *Nitric Oxide* **24** 102–9
Rinard G A, Quine R W, Eaton S S and Eaton G R 2002 Frequency dependence of EPR signal intensity, 250 MHz to 9.1 GHz *J. Magn. Reson.* **156** 113–21
Romanyukha A, Trompier F and Reyes R A 2014 Q-band electron paramagnetic resonance dosimetry in tooth enamel: biopsy procedure and determination of dose detection limit *Radiat. Environ. Biophys.* **53** 305–10
Satpathy A, Ranjan R, Priyadarsini S, Gupta S, Mathur P and Mishra M 2019 Diagnostic imaging techniques in oral diseases *Medical Imaging Methods. Recent Trends* ed A K Shukla (Singapore: Springer Nature) ch 3 pp 59–96
Schröder H E 2000 *Orale Strukturbiologie* 5th edn (Stuttgart: Thieme)
Shah J S, Shah H A and Rathva Y 2021 Nodular malignant melanoma of maxillary gingivae: a rare presentation and a proposed clinical classification of oral malignant melanoma *J. Indian Acad. Oral Med. Radiol.* **33** 107–10
Skaleric U, Cevc P, Gaspirc B and Schara M 1998 Crystallite arrangement of hydroxyapatite microcrystals in human tooth cementum as revealed by electron paramagnetic resonance (EPR) *Eur. J. Oral Sci.* **106** 365–7

Slominski R A, Żmijewski M and Slominski A T 2015 The role of melanin pigment in melanoma *Exp. Dermatol.* **24** 258–9

Soares A B, de Araujo V C, Passador-Santos F, Thomaz L A, de Freitas A L S, Mautoni M C, Stelini R F and Cintra M L 2021 Uncommon pigmented carcinoma *in situ*: case report and brief review *Clin. Pathol.* **14** 2632010X211009819

Sroka A, Sztukowska M, Potempa J and Travis J 2001 Genco CADegradation of host heme proteins by lysine- and arginine-specific cysteine proteinases (gingipains) of *Porphyromonas gingivalis J. Bacteriol.* **183** 5609–16

Strancar J and Petelin M 2010 Simulation support to EPR-based dentin characterization *Appl. Magnet. Reson.* **37** 751–61

Summitt J B, Robbins J W, Hilton T J and Schwartz R S 2006 *Fundamentals of Operative Dentistry: A Contemporary Approach* (Hanover Park, IL: Quintessence)

Takahama U, Hirot S and Takayuki O 2008 Detection of nitric oxide and its derivatives in human mixed saliva and acidified saliva *Meth. Enzymol.* **440** 381–96

Uhlenbeck G E and Goudsmit S 1926 Spinning electrons and the structure of spectra *Nature* **117** 264–5

Umakoshi M, Yamaguchi I, Hirata H, Kunugita N, Williams B B, Swartz H M and Miyake M 2017 *In vivo* electron paramagnetic resonance tooth dosimetry: dependence of radiation-induced signal amplitude on the enamel thickness and surface area of *ex vivo* human teeth *Health Phys* **113** 262–70

Vasileiou T and Summerer L 2020 A biomimetic approach to shielding from ionizing radiation: the case of melanized fungi *PLoS One* **15** e0229921

Vicentin B L S, Peron R F, Amorim B C, Buelvas D D A, Di Mauro E, Borges H O L and Hoeppner M G 2019 Application of ESR spectroscopy in the study of the light efficiency of a dental translucent fiberglass post to promote polymerization *J. Appl. Spectrosc.* **86** 244–9

Vicentin B L S and Di Mauro E 2019 Free radicals and polymerization of resinous materials used in dentistry *Electron Spin Resonance in Medicine* ed A K Shukla (Singapore: Springer Nature) ch 6 pp 103–34

Vicentin B L S, Netto A M, Blumich B and Di Mauro E 2016 Identification of free radicals generated by different curing modes in a dental resin cement *Appl. Magnet. Reson.* **47** 1003–14

Wang Y, Huang X and He F 2019 Mechanism and role of nitric oxide signaling in periodontitis *Exp. Ther. Med.* **18** 3929–35

Wertz J E and Bolton J R 1986 *Electron Spin Resonance: Elementary Theory and Practical Applications* (London: Chapman and Hall)

Wolf H E D, Edith M, Rateitschak H K and Hassel T M 2005 *Periodontology* 3rd edn (Stuttgart: Thieme)

Yoshino F, Yoshida A, Wada-Takahashi S, Sugiyama S, Tokutomi F, Maehata Y, Miyamoto C, Komatsu T, Takahashi S S and Kobayashi K 2012 Assessments of salivary antioxidant activity using electron spin resonance spectroscopy *Arch. Oral Biol.* **57** 654–62

Zdravkova M, Wieser A, El-Faramawy N, Ivanov D, Gallez B and Debuyst R 2002 An *in vitro* L-band EPR study with whole human teeth in a surface coil resonator *Radiat. Meas.* **37** 347–53

Zeeman P 1897 The effect of magnetisation on the nature of light emitted by a substance *Nature* **55** 347

Chapter 6

The application of molecular spectroscopy and chemometrics in dentistry

Gunawan Indrayanto and Abdul Rohman[1]

6.1 Introduction

Recently, analytical methods based on molecular spectroscopy (MS) have gained popularity in the analysis of health-related issues, including in dentistry. Analytical methods based on MS provide numerous advantages, including simplicity in sample preparation, non-invasiveness, rapidity, minimal use of solvents and reagents (which can be described as green analytical chemistry), and its capability for monitoring several analytes in the evaluated samples [1]. The term MS applies to to any analytical methods relying on the interaction between analytes and electromagnetic radiation (EMR), expressed as wavelength (UV–vis spectroscopy) or frequencies/wavenumbers (infrared and Raman spectroscopy) at the molecular level. The use of the term 'molecular' in MS is intended to contrast with 'atomic' in atomic spectroscopy. MS analyzes analytes in molecular states, while analytes are analyzed at the atomic level during analysis using atomic spectroscopy [2]. The different types and compositions of certain analytes, such as the minerals present in dental components and early detection of dental caries (one of the most prevalent diseases affecting human teeth at all ages), and other characterizations of certain compounds in enamel, dentin or whole teeth, can be studied using MS [3].

The spectra resulting from of MS are quite complex, therefore the molecular spectra, in particular infrared and NMR spectra, are quite difficult to evaluate. Fortunately, with the advanced development of statistical software, chemometrics applying statistical and mathematical calculations has proven very beneficial for extracting more information from the molecular spectrum data. In analytical chemistry and big data management, various chemometric techniques have been

[1] Center of Excellence, Institute for Halal Industry and System (IHIS), Gadjah Mada University, Yogyakarta 55281, Indonesia, **abdulkimfar@gmail.com**.

doi:10.1088/978-0-7503-3459-4ch6

applied for the treatment of chemical data to be more understandable, not only for chemists but also non-chemists. Therefore, it is not surprising that chemometrics has emerged as a valuable tool in several fields such as medicine, pharmaceuticals, environment, agriculture and dentistry [4]. In the analysis of dental materials and the early detection of dental-related diseases, chemometrics is widely used to treat the data, including pre-processing data, the chemometrics of classification and multivariate calibrations.

6.2 Molecular spectroscopy

MS is concerned with the interactions between molecules (analytes in the molecular form) and EMR, expressed as wavelength or frequency. Molecules in the analyzed samples can absorb or emit EMR to result in specific phenomena. In UV–vis spectra, a species absorbing UV radiation undergoes an electronic transition from the ground state to an excited state. A species emitting radiation undergoes a transition from a higher energy excited state to a lower energy state [5]. In addition, functional groups absorbing IR radiation can result in vibrational transitions. The types of EMR with their wavenumbers $\tilde{\nu}$ (cm^{-1}) are: ultraviolet–visible ($>14\ 000\ cm^{-1}$, corresponding to electronic transitions), near-infrared (14 000–4000 cm^{-1}, molecular vibrations), mid-infrared (4000–400 cm^{-1}, molecular vibrations), far infrared (400–4 cm^{-1}, molecular vibration), and microwaves ($<4\ cm^{-1}$, molecular rotations). If the wavenumber is $\tilde{\nu}\ cm^{-1}$, the wavelength λ (nm) is $\frac{10^7}{\tilde{\nu}}$ [6].

This section will summarize the basic principles, methods of sample preparation, applications and method validation of some techniques of MS, that are described by general chapters of the current *United States Pharmacopeia* (USP) and *National Formulary* (NF), USP 43–NF 38. The general chapters which will be discussed are as follows: <857> 'Ultraviolet–Visible Spectroscopy' [7], <1857> 'Ultraviolet–Visible Spectroscopy, Theory and Practice' [8], <853> 'Florescence Spectroscopy' [9], <1853> 'Florescence Spectroscopy, Theory and Practice' [10], <856> 'Near-Infrared Spectroscopy (NIRS)' [11], <1856> 'NIRS, Theory and Practice' [12], <854> 'Mid Infrared Spectroscopy (MIRS)' [13], <1854> 'Mid Infrared, Theory and Practice' [14], <858> 'Raman Spectroscopy (RS)' [15], <1858> 'Raman Spectroscopy, Theory and Practice' [16], and other related general chapters of the USP 43–NF 38 and other publications will be also included.

For ensuring that the analytical instrument is suitable and meets the specification, an analytical instrument qualification should be first performed. The qualification is comprised of three steps: installation qualification (IQ), operational qualification (OQ), and performance qualification (PQ). General methods for instrument qualifications can be found in the general chapter 1058 on 'Analytical Instrument Qualification' of the current USP 43–NF 38 [17]. The principles of instrumentation and the qualification of different types of spectrophotometers are described in the respective general chapters 'Ultraviolet–Visible Spectrophotometer' [7], 'Florescence Spectrophotometer' [9], 'Near-Infrared Spectrophotometer' [11], 'Mid-Infrared Spectrophotometer' [13] and 'Raman Spectrophotometer' (<858>) [15]. The most important parameter that must be evaluated for OQ is the wavelength accuracy. All spectrophotometers must meet the acceptance criteria of IQ, OQ and PQ, as required by the respective chapters. For the acceptance criteria of analytical method validation, as discussed in this section for drugs and related products, the readers can refer to other current formal guidance for food samples, and for research purposes self-pre-determined criteria can be applied.

6.2.1 Ultraviolet–visible spectroscopy

6.2.1.1 Principles of measurement

UV–vis spectra are the result from interaction of EMR at wavelengths of 200–780 nm with the electron cloud in the compounds having chromophoric groups. During the interaction, the electronic transition of one or more electrons from a ground state into a higher energy state occurs. The UV–vis spectral bands of compounds are typically broad and are not specific to a certain compound. The Beer–Lambert law states that the absorbance (A_λ) of a solution at λ nm is defined as the logarithm of the reciprocal of the transmittance (T_λ) as described by

$$A_\lambda = \log\frac{1}{T_\lambda} \text{ and } T_\lambda = \frac{I_\lambda}{I_{\lambda 0}}, \tag{6.1}$$

where T_λ is transmittance, I_λ is the intensity of the transmitted radiation at the same wavelength λ, and $I_{\lambda 0}$ is the intensity of the incident radiation at wavelength λ. The absorbance (A) is proportional to the path length (b) and to the concentration (c) of the compound in the solution:

$$A = \varepsilon_\lambda .\ c.\ b, \tag{6.2}$$

where $\varepsilon\lambda$ is the molar absorptivity, c is the compound concentration (g/L) and b is the path length (cm). If the concentration, c, is expressed in g/L, the constant ε_λ becomes a_λ, which is called the absorptivity. The specific absorbance of a dissolved compound, referring to the absorbance of a 1% weight solution in a 1 cm cell measured at λ, can be calculated using

$$A_{1\ \text{cm}}^{1\%} = 10a_\lambda = 10\frac{\varepsilon_\lambda}{M}, \tag{6.3}$$

where M is the molar concentration of the solution. Solutions (10 μg mL^{-1}) observed in 1 cm cells will produce absorbances of ca. 0.2–0.8 in the UV or visible region [7].

UV–vis spectra can be transformed into their first-order or higher (second- and higher order) derivative spectra. First-order derivatives can be applied for minimization or elimination of the background in turbid and scattering solutions and suspensions and for analysis of trace components in complex absorbing matrices. Second-order or higher can be used for (i) enhancing the resolution of overlapping peaks for the separation of superimposed spectra, which is particularly useful in multicomponent analysis, (ii) assisting in the quantitative analysis of trace components, (iii) assisting in the the characterization of individual pure compounds, particularly for archiving purposes and for complementing the information obtained from other techniques such as infrared, nuclear magnetic resonance, and mass spectroscopy, and (iv) assisting in purity testing of dental products [8].

6.2.1.2 Sample solution preparations

For preparation of samples for UV–vis spectroscopic measurements, the samples are typically is dissolved in a certain solvent. Unless otherwise directed in the monograph in the Pharmacopoeia, UV–vis spectral measurements are usually carried out at ambient temperature using a cell with a path length of 1 cm. Many solvents are suitable to be used in UV–vis spectroscopy, including water, alcohols, acetone, lower hydrocarbons such as hexane, ethers, and dilute solutions of strong acids and alkalis. The solvents employed for UV-spectral measurement must be free from any contaminants capable of absorbing in the spectral region used. Water must be free with methanol or ethanol or alcohol denatured by the addition of methanol but without benzene or other interfering impurities. It is recommended to use spectrometric quality solvents, which are ensured to be free from any contaminants. New lots of the solvents used should be checked for their transparency, and analysts should use the same lot of solvent for preparation of the test solution, the standard solution, and the blank. The best practice is to use solvents that have at least 40% transmittance (39.9% $T = 0.399$ A) at the wavelength of interest [7].

6.2.1.3 Method validation

An analytical method validation procedure is required when a new method is proposed to be used as an alternative method to the official procedure of the Pharmacopeia, or when no official procedure exists. The objective of method validation is to establish that the proposed method is suitable for its intended purpose, including the quantitative analysis of the main component in a drug substance or a drug product (category I assays), quantitative analysis of impurities or limit tests (category II), and identification tests (category IV), as described by USP 43–NF 38, general chapter <1225> 'Validation of Compendial Procedures', table 2 [7, 18].

Specificity test. Specificity is ensured using a reference standard wherever possible; it is demonstrated by the lack of interference from other components or impurities present in the matrix [7].

Linearity evaluation. Linearity evaluation is related to the linear relationship between the concentration of the analytes and the UV–vis response. The linearity must be verified using at least five solutions of working standards at different concentrations around the expected concentration of the tested solution. The performance of the standard curve is further evaluated using appropriate statistical methods including a least-squares regression, and the Pearson correlation coefficient r is calculated along with other statistical parameters. The coefficient of determination, r^2, must be at least 0.995 for category I assays and at least 0.99 for category II quantitative tests. Visual inspection of the residual plots should not disclose any significant pattern. The value of r^2 can be used as a measure of the fraction of the data's variation that is adequately modeled and not a measure of linearity. Linearity depends on the standard error of the calibration equation and on the range of the calibration data. Thus, although values very near 1.00, such as 0.99 or greater, typically indicate a linear relationship, lower values do not distinguish between nonlinearity and variability [7]. Therefore, to prove the linearity of the calibration curve other parameters must be evaluated. Detailed discussion regarding the evaluation of the linearity can be found in a previous publication [19].

Accuracy. For categories I, II and III, the accuracy procedures can be evaluated using recovery studies with the appropriate matrix. During recovery studies the samples are spiked with known concentrations of the analyte. Analysts also can compare assay results obtained using the UV–vis procedure under validation to those from an established analytical procedure or reference standard method. Acceptance criteria: A mean recovery of 98.0%–102.0% is accepted for drug substances, a mean recovery of 95.0%–105.0% is accepted for drug product assays and the acceptable mean recovery for impurity analysis is 80.0%–120.0%. These criteria are to be met throughout the intended range [7]. If the accuracy is evaluated by comparing to an established validated procedure, the readers should consult the methods that are described by USP 43–NF 38 general chapter <1010> 'Analytical Data—Interpretation and Treatment' [20].

Repeatability. Repeatability is evaluated by analyzing six concentrations of prepared sample solutions independently at 100% of the expected concentration. Otherwise, it can be determined by analyzing the concentrations of three replicates of three separate sample solutions at different concentrations. Acceptance criteria: The RSD should be at least 1.0% (drug substance assay), 2.0% (drug product assay) and 15.0%–20.0% (impurity analysis) [7].

Intermediate precision. The effect of random variables on the analytical precision of the method must be evaluated. Typical variables are evaluated by carrying out the analysis on different days, using different analytical instruments and by two or more analysts. Any combination of at least two of these variables totaling six experiments will provide an assessment of intermediate precision. Acceptance criteria: The RSD values should be at least 1.5% (drug substance assay), 3.0% (drug product assay) and 15.0%–25.0% (impurity analysis) [7].

Detection limit (DL) and quantification limit (QL). The DL and QL can be determined by calculating the SD of at least six replicate measurements of a blank solution and multiplying by 3.3 (DL) or 10 (QL). Otherwise, the SD can be analyzed from the intercept error from a calibration curve, or by determining the ratio of signal-to-noise (S/N), which should be >3.3 (DL) or 10 (QL). The estimated DL must be confirmed by analyzing samples at the calculated DL. For confirming sufficient sensitivity and an adequate precision of QL, a representative sample matrix spiked with analyte at the QL concentration is analyzed. The observed signal-to-noise ratio at the required QL should be >10. QL is considered valid if the measured concentration is accurate and precise at a level ⩽50% of the specification [7].

Range. The range of an analytical procedure is the interval between the upper and lower concentrations of analyte in the sample, for which it shows suitable level of precision, accuracy and linearity. Acceptance criteria: for category I tests, the validation range is 80.0%–120.0% (for 100.0% centered acceptance criteria), or 10.0% below the lower limit to 10.0% above the upper limit (for non-centered acceptance criteria), and 70.0%–130.0% (for content uniformity). For category II tests the validation range covers 50.0%–120.0% of the acceptance criteria [7].

Robustness. The robustness parameter is evaluated during analytical method development. For UV–vis, the robustness can be evaluated by measuring the stability of the analyte under specified storage conditions, varying pH, and adding possible interfering species in the samples [7].

6.2.2 Fluorescence spectroscopy (FS)

6.2.2.1 Principles of measurement

FS is a two-step process which requires the absorption of light or EMR at a specific wavelength called the excitation wavelength (λ_{ex}) followed by emission of light, generally at a higher wavelength, called the emission wavelength (λ_{em}). The emission of light is called fluorescence. The absorption of a photon by a molecule in the sample promotes an electron to be excited into an excited state with higher energy than the ground state; the excited electron returns to the ground electronic state by emitting a photon. If the emission arises from an 'allowed' transition that typically has a short lifetime (1–10 ns), then it is called fluorescence. If the emission comes from a 'forbidden' transition that typically has a long lifetime (1 ms–1 s), then it is called phosphorescence. Phosphorescence is usually less intense than fluorescence. Phosphorescence typically has a higher emission wavelength than fluorescence. Fluorescence techniques are also background-free due to the low excitation of light that reaches the detector. This typically makes fluorescence detection highly sensitive, down to single-molecule detection in some cases. Fluorescence detection can also be highly specific because a fluorophore emits a characteristic emission pattern. The two performance characteristics of specificity and sensitivity are the more important strengths of fluorescence methods. In contrast to absorption spectroscopy, where deviation from linearity is the exception, fluorescence linearity can be affected by several sample-related effects [9, 10].

The common type of sample to be analyzed using fluorescence techniques is a diluted, transparent solution which absorbs light following the Beer–Lambert law and that emits a corresponding fluorescence intensity that is directly related to the concentration, the absorptivity and the fluorescence quantum yield of the fluorescent species or fluorophore. If the excitation wavelength is fixed and the emission wavelength is measured, the resultant spectrum is called an emission spectrum. If the emission wavelength is fixed and the excitation wavelength is measured, the resulting spectrum is called an excitation spectrum. The form of a fluorescence spectrum is like a UV–vis–NIR absorption spectrum [10].

6.2.2.2 Sample solution preparation, and qualitative and quantitative measurements

The sample solution can be prepared as described in section 6.2.1.2. Solvents can be used if an interfering fluorescence signature at the wavelength(s) of interest was not observed. The fluorescence baseline intensity should not be <2% of the expected measurement signal unless a larger value has been justified previously [9]. Direct sample measurements can be performed for the analysis of dental applications such as tissues and resin composite [21].

Qualitative fluorescence methods are used to spot the occurrence of analytes and yield a positive or negative result. The excitation (λ_{ex}) and emission wavelengths

(λ_{em}) are frequently selected at the peak maximum of the fluorophore. The minimum amount of analyte needed for a positive result should be considered by the analyst to confirm that the method is suitable for the application. The observation of fluorescence at the peak position above the limit of detection (usually three times the noise level) indicates a positive result [9].

Quantitative fluorescence methods are used to determine the concentrations of analytes in samples. Concentration may be determined as absolute units (moles or moles L^{-1}) or in relative units, i.e. the ratio of the concentrations of two fluorescent analytes contained in a certain sample. The fluorescence signal at a given pair of excitation (λ_{ex}) and emission wavelengths (λ_{em}) is correlated to fluorescent analyte concentration. This linear correlation is only valid for optically dilute samples (e.g. solutions with an absorbance of less than 0.05 at a path length of 1 cm) [9]. The calibration curves of fluorescence intensity versus analyte concentration are determined using a reference standard compound. The calibration curve is both analyte and instrument specific and can be used to determine the analyte concentrations of samples. The calibration curve maybe inaccurate if the microenvironment surrounding the fluorophore is different in the reference and the sample [10].

There are two main factors that can affect the results of the quantification, i.e. instrument-based and sample-based factors. Instrument parameters such as wavelengths, bandwidths and detector gain can affect the results of quantification. All these parameters can be set with varying degrees of repeatability and accuracy, depending on the instrument used. As an example, the measured peak positions of the emission bands of two analytes may differ between instruments because of a wavelength bias. The intensity of the excitation beam can change significantly with excitation wavelength or with time because of the wavelength dependence of the intensity of the light source and the transmittance of the excitation wavelength selector or the time dependence of the light source intensity. Changes of instrumental polarization settings can change the observed excitation intensity and the response of the detection system. The response of the detector is not linear with intensity at all intensities, so analysts should know the linear intensity range of the detection system used. Analysts must determine the linear range of the fluorescence detection system before attempting to calibrate the detector responses [10].

The fluorescence intensity of concentrated samples (e.g. absorbance $A > 0.05$ at a path length of 1 cm) does not increase linearly with concentration because of significant absorption of the excitation beam and/or emission (reabsorption) by the sample. The fluorescence intensity of a sample may decrease with time of exposure to light because of photo-bleaching and photo-degradation, as shown by most organic dyes, which are the most widely used fluorescent probes. Analysts should limit the time of exposure to the light for those samples in order to obtain reproducible fluorescence intensities and spectral shapes. The fluorescence intensity of the fluorophores is also temperature dependent. The fluorescence intensity, peak position and sometimes even the spectral shape are often dependent on the environment, including the solvent used, the solution's pH, or the species to which the fluorophore is bound [10].

6.2.2.3 Method validation

The methods of evaluation and acceptance criteria of specificity, linearity, accuracy, DL, QL and range are identical to those in section 6.2.1.3. The methods for repeatability and intermediate precision are identical to those in section 6.2.1.3. The acceptance criteria for repeatability are: RSD is at least 1.0% (for a drug substance), 2.0% (for a drug product assay) and 20.0% (for impurity analysis); and for intermediate precision are: RSD is at least 1.0% (for a drug substance), 3.0% (for a drug product assay) and 25.0% (for impurity analysis).

Parameters that should be evaluated for robustness include measuring the stability of the analyte under specified storage conditions, varying pH, removal of oxygen and adding possible interfering species [9].

6.2.3 Near-infrared spectroscopy (NIRS)

6.2.3.1 Principles of measurement

NIRS is a vibrational spectroscopic method that measures the absorption of EMR at 780 to 2500 nm (ca. 12 821–4000 cm^{-1}), which is resonant with frequencies from molecular dipole changes. The NIR spectroscopic region is dominated by overtone and combination frequencies of the molecular fundamental harmonic frequencies in the mid-IR region. The molar absorptivity in the NIR region is much lower than those in the IR region, and the radiation can penetrate several millimeters into the material, giving rise to a wavelength-dependent scattering signal overlapping the molecular absorbance signals. Materials such as glass are relatively NIR transparent. NIR spectroscopy is useful in the area of pharmaceutical analysis for either qualitative or quantitative applications [11]. NIR spectra are dominated by C–H, N–H, O–H and S–H overtone resonances and combinations of fundamental mid-infrared vibrational modes from which they originate [12]. The most common measurement methods of NIR spectroscopy are transmission and reflection. Incident NIR radiation is absorbed or scattered by the sample and is measured as transmittance or reflectance, respectively [12].

Transmittance (T) is the intensity ratio of the transmitted radiation (I) to the incident radiation (I_0), which represents the decrease in intensity at given wavelengths when radiation is passed through the sample; NIR spectra are usually measured in absorbance (A):

$$T = \frac{I}{I_0}, \text{ or } T = 10^{-A}. \tag{6.4}$$

Transmittance (T) is dependent on a background transmittance spectrum for its calculation. Examples of background references include air, a polymeric disc, an empty cell, a solvent blank, or, in special cases, a reference sample. The method usually applies to gases, liquids (diluted or undiluted), dispersions, solutions and solids, including tablets and capsules [12].

Reflectance (R) is the ratio of the intensity of light reflected from the sample, I, to that reflected from a background or reference reflective surface, I_r:

$$R = \frac{I}{I_r} \tag{6.5}$$

$$A = \log \frac{1}{R}. \tag{6.6}$$

The NIR reflection mode is usually used for scattering samples such as powders and slurries. NIR reflectance spectra are generally evaluated by calculating and plotting $\log(1/R)$ versus wavelength. This log form is the pseudo-absorbance of the material and is commonly called absorbance [12].

The diffuse reflection (DR) mode gives a measure of reflectance (R) (penetrating the sample surface, interacting with the sample, and passing back through the sample's surface), which is the ratio of the intensity of light diffused and reflected from the sample (I) to that reflected from a background or reference reflective surface (I_r). R can be calculated using equation (6.4), I_r is used instead I. This method generally applies to solids [12].

Transflection spectrometry (T*) is a hybrid of transmission and reflection wherein a reflector is placed behind the sample so that the optical path through the sample and back to the detector is doubled, compared with a transmission measurement of a sample of the same thickness. Non-absorbed radiation is reflected from the sample to the detector. The light may be reflected from a diffuse or specular (mirror) reflector placed behind the sample.

$$T^* = \frac{I}{I_{T*}}, \tag{6.7}$$

in which I is the intensity of transflected radiation measured with the sample, and I_{T*} is the intensity of transflected radiation of the reference material as the background. This method generally applies to liquids and clear plastic materials [12].

6.2.3.2 Methods of measurement and applications

NIR spectra can be obtained from several sample presentations, including solids, powders, slurries, gels, liquids, films and gases. In addition, the measurement can be performed through glass or plastic films that are normally used for containment. Typically, NIR spectroscopy does not require any sample preparation because the detectors have a wide dynamic range, and the collection time can be adjusted by the user. Two factors, environmental (humidity and temperature) and sampling factors (e.g. sample presentation area, solvent, sample thickness, solid state forms, etc), can affect the NIR spectra [11, 12].

NIRS can be applied for several purposes: (i) identification of active substances, excipients, dosage forms, manufacturing intermediates, chemical absorbance materials and packaging materials; (ii) qualification of ingredients, intermediates, and drug products, including batch-to-batch spectral comparison and supplier change assessment; (iii) quantification of active substances and excipients, determination of chemical values such as hydroxyl values, determination of absolute water content, determination of the degree of hydroxylation and control of solvent content;

(iv) physical characterizations including crystalline form and crystallinity, polymorphism, solvates and particle size, and analysis of intact pharmaceutical dosage forms such as tablets, capsules, lyophilized products and implants (e.g. polymeric and microspheres), dissolution behavior, disintegration patterns, hardness, and examination of film properties; and (v) in-line processing monitoring, such as drying, granulation, coating, blending, crystallization, etc, as an example NIRS can be used for determining the endpoint of a blending operation [12].

Diffuse reflectance near-infrared spectroscopy in combination with chemometrics of *k*-nearest neighbor method (kNN) and neural network (NN) has been used for detection of dental caries and for the classification between healthy tissue and caries lesions. All data were generated from near-infrared system equipped with InGaAs detector. The results revealed that kNN and NN could distinguish the healthy tissue and caries lesions with the accuracy level of 94.1%–98.4% with the error value during cross-validation was lower than 8.3% [22].

6.2.3.3 Method validation

Specificity. The extent of specificity testing depends on the intended application. Specificity is typically demonstrated by using the following approaches. Qualitative analysis and identification or confirmation testing are the common applications of qualitative NIRS. Identification is achieved by comparing a sample spectrum to the reference spectrum or spectra, for example by establishing a mathematical relationship (chemometrics model). The specificity of the NIRS identification procedure is demonstrated by obtaining positive identification from samples, coupled with negative results from materials that should not meet the criteria for positive identification. Typically, quantitative applications of NIRS establish a mathematical relationship (chemometrics model) between the NIR spectral response and physical or chemical property of interest. The demonstration of specificity may include the following: spectral regions can be correlated to the property of interest; wavelengths can be used by regression analysis for calibration; variation in spectra from samples for calibration should be examined and interpreted; and variation in the material composition and sample matrix may be shown to have no significant effect on quantification of the property of interest within the specified procedure range [11].

Linearity. A linear relationship between the NIR spectral response function and the property of interest (typically concentration). Although demonstrating a linear response is not required for all NIR applications, the model chosen should properly represent the relationship. The number of validation samples for calibration will depend on their complexity and interferences (e.g. temperature, particle size, and others); critical interferences must be identified. The results of quantitative determination must be within the calibration interval. Multiple linear regression, principal component regression and partial least-squares regression are generally used [11, 12].

Robustness. Typical measurement parameters for robustness include the effects of environmental conditions (e.g. temperature, humidity, and vibration), sample temperature, sample handling (e.g. fiber-optic probe depth, compression of material, sample depth/thickness) and instrument changes (e.g. lamp change, warm-up time) [11].

The methods of evaluation and acceptance criteria for accuracy, repeatability and intermediate precision, and the acceptance criteria for linearity are identical to those in section 1.2.3 [11]. For methods of analysis that use chemometrics, see the general chapter <1039> [23].

6.2.4 Mid-infrared spectroscopy (MIRS)

6.2.4.1 Principles of measurement

MIRS comprises measurement of the absorption of EMR over the wavenumber range of 4000–400 cm^{-1} (that corresponds to the wavelength range of 2.5–25 μm) caused by the promotion of molecules from the ground state of their vibrational modes to an excited vibrational state. Fundamental transitions from the ground state of a given mode ($v_i = 0$) to its first excited vibrational state ($v_i = 1$) yields the strongest bands, though weaker overtone and combination bands also are observed in the spectrum. The promotion of molecules from their ground state to their second and higher vibrational states ($v_i = 2$, 3, etc) gives rise overtone bands. Vibrational modes include the motion of all atoms of the molecule. If molecules contain a certain functional group, the transitions often arise in narrow spectral ranges. In this case the wavenumbers at which these transitions occur are known as group frequencies. If a vibrational mode comprises atomic motions of more than just a few atoms, the frequencies occur over wider spectral ranges and are not characteristic of a functional group, they are more characteristic of the molecules. Such bands are known as fingerprint bands [14]. IR spectroscopy offers a specific method for the identification of analyte(s) due to its property as a fingerprint technique.

All strong bands absorbing IR radiation at wavenumbers above 1500 cm^{-1} are group frequencies, whilst those that absorb below 1500 cm^{-1} can either be group frequencies or fingerprint bands. Therefore not all strong bands in the IR spectrum of a given molecule can be attributed to the presence of a functional group. Polar groups, such as C–O, C=O, O–H, N–H and C–F, typically yield strong bands. Functional XY_2 (such as $-CH_2$, $-NH_2$, $-NO_2$ and $-SO_2$) and XY_3 (such as $-CH_3$ and $-NH_3^+$) will show first-order coupling, so that the mode is split into a symmetrical mode (where both the Y atoms move to and from the X atom with the same phase) and an anti-symmetrical mode (where one Y atom moves 180° out of phase with the other). The amount by which the two bands are split depends on the Y–X–Y angle [14]. Some functional groups represented by peaks and shoulders in IR spectra are used as markers during analysis. For example, the functional groups of amide bands I, II and III of collagen, the phosphate band and the carbonate band are used in the characterization of dentin. The absorbance peaks of amide I and ν_3 PO_4 has been used for the decalcification of dentin surfaces after acid treatment [24].

6.2.4.2 Procedure of measurement and applications

Mid-IR spectra can be measured by transmission, external reflection, internal reflection (often called attenuated total reflection (ATR)), DR and microscopy. Different techniques for the sample preparation are available for these options. The most common sample preparation techniques are potassium bromide (KBr) disks, mineral oil mulls, self-supported polymer films, capillary films, liquids and solutions in transmission cells, gases, ATR, diffuse reflection and microscope sampling [13, 14].

USP 43–NF 38 general chapter <197> describes several methods for the preparation of test samples [25]:

⟨197A⟩: The substance under examination is intimately in contact with an internal reflection element for ATR analysis.

<197D>: The substance under examination is mixed intimately with potassium bromide and transferred to a sample container for DR analysis.

<197E>: The substance under examination is pressed as a thin sample against a suitable plate for IR microscopic analysis.

<197F>: The substance under examination is a thin film of a neat liquid or semisolid between suitable (e.g. sodium chloride or potassium bromide) plates, or a thin microcrystalline or glassy film deposited from a solution or cooled from a melt.

<197K>: The substance under examination is mixed intimately with potassium bromide and compressed into a transparent pellet.

<197M>: The substance under examination (10–20 mg) is finely ground and dispersed in a drop of mulling agent (unless otherwise directed in the monograph, mineral oil is to be used).

<197S>: A solution of designated concentration is prepared in the solvent specified in the individual monograph. The solution is examined in 0.1 mm cells, unless a different cell path length is specified.

Interested readers can consult our previous chapter regarding detailed sampling techniques for MIRS [26]. Infrared spectroscopy using the mid region of 4000–400 cm^{-1} has been used successfully for fast differentiation of typical dental materials produced from organic compounds for inorganic restorations (porcelain) and for characterization of dental composites [27]. In addition, MIR spectroscopy is capable of identifying tooth structures which resemble hydroxyapatite and its contamination with ions of fluoride and carbonate [28].

6.2.4.3 Method validation

The methods and acceptance criteria for linearity, accuracy, reputability, intermediate precision, DL, QL and range are identical to those in section 1.2.3.

Specificity. For the identification test, confirmation is needed using positive (using reference material) and negative controls (sample without target compound). For category IV tests the identity of the analyte should be ensured. For category I and II procedures, the accuracy requirement also demonstrates specificity for the targeted analytes.

Robustness. Changes of sample preparation methods and hardware settings should be evaluated [13]. For analysis that uses chemometrics see also the general chapter <1039> [23].

6.2.5 Raman spectroscopy (RS)

6.2.5.1 Principles of measurement

RS is a vibrational spectroscopic technique. RS is complementary to MIRS and NIRS. The principle of RS is the excitation of the studied samples with a laser of monochromatic light in the visible spectral range. The relative intensity of bands in RS could be related to the concentration of the studied samples which can be used for quantitative analysis according to the Lambert–Beer law [29]. The Raman effect arises because of a change in the polarizability of molecular bonds through a given vibrational mode and is measured as inelastically scattered radiation, while the absorption intensity at the IR or NIR wavelengths occurs due to differences in dipole moments. RS is mainly sensitive to non-polar bonds (e.g. C–C, single or multiple bonds) and thus strong dipole oscillators do not display Raman scattering because the molecular geometry does not give rise to polarizability. The positions of the (Raman-shifted) wavenumbers for a given vibrational mode are theoretically equal to the wavenumbers of the corresponding bands in an IR absorption spectrum. However, the stronger peaks in a Raman spectrum are often weak, or not present in an IR spectrum, and vice versa. Likewise, wavenumbers above 1500 cm^{-1} are called group frequencies, and strong bands that absorb below 1500 cm^{-1} can be either group frequencies or fingerprint bands for the molecule. Water, which has a strong IR absorption spectrum, only exhibits a weak Raman peak and is a suitable solvent for this technique. The spectra of Raman and IR/NIR are different for most organic molecules, and thus the spectra of these different techniques can be applied to complement specificity and characterization [15, 16].

A Raman spectrum is produced by exciting the sample to an excited state with a suitable light source. Elastically scattered light is known as Rayleigh scattering and is not of interest in RS, except for marking the laser wavelength. If the sample relaxes to a vibrational energy level that differs from the initial state either up or down, the scattered radiation is shifted in energy. This shift is equal to the energy difference between the initial and final vibrational states. This 'inelastically scattered' light is mentioned as Raman scatter. Only around 1 of the 106–108 photons incident on the sample undergoes Raman scattering. A laser that delivers high photon densities is used in RS for compensation of the low Raman scattering efficiency. The Rayleigh scatter can be filtered for reducing stray light interferences. Lasers can be tuned for avoiding fluorescence of the sample or matrix or for enhancing the signals. Detailed discussions regarding this issue are available in [16].

6.2.5.2 Methods of measurement and applications

Raman spectra can be acquired from many kinds of samples including solids, powders, slurries, gels, liquids, films and gases. Measurements can be made through the containers, e.g. glass or plastic film. Characteristically, RS does not require any sample preparation due to its relative insensitivity to the physical state (e.g. solid, emulsion, or slurry). Raman uses backscattering geometry, right angle geometry, or transmission geometry, so the frequency and intensity are not affected much. Any intensity difference may arise from non-homogeneous samples and different sampling areas [15].

RS can be applied for the analysis of incoming raw materials, process monitoring of drug production, and analyzing finished products. Other applications of RS are optimizing polymorph analysis, the synthesis of new drugs, gaining process understanding through real-time monitoring of reactors, crystallizers and blenders, analyzing raw materials and ensuring product quality. For drug synthesis, RS is ideal for monitoring reactant, intermediate and product concentrations, as well as determining the pathways, kinetics, mechanisms, endpoints and yields for a variety of reaction types. RS can be used in the quality-by-design control of a product, such that finished products have the correct amount of active pharmaceutical ingredient, its polymorph, excipient, and other additives such as dyes. RS has been used for monitoring mixing in blenders, as well as to inspect individual products before shipment. RS can be applied for stability studies as an alternative to HPLC. In dentistry applications, RS is reported for the identification of tissue changes of teeth due to periodontitis [30], the early detection of dental caries, and monitoring the demineralization of dentin, which represents an important issue in modern dentistry [31]. RS has also been used for the detection of some trace elements present in the enamel [32], for detecting the formation of dentin in pulp tissue [33] and for the identification of caries at an early stage before visual changes in enamel [34]. Chemometric tools are common methods for evaluating the described applications [16].

6.2.5.3 Method validation

The methods and acceptance criteria for accuracy, reputability, intermediate precision and range are identical to those in section 1.2.3. For analysis that used chemometrics, see the general chapter <1039> [23].

Linearity and calibration model. The linearity and calibration model involves demonstrating a linear relationship between the Raman spectral response function and the property of interest. Although demonstrating a linear response is not required for all Raman applications, the model chosen should properly represent the relationship. Validation of linearity in Raman procedures can be performed by measuring at least five samples in the expected concentration range and by plotting either the Raman spectral responses versus actual or accepted values for the property of interest. Linearity depends on the standard error of the calibration function and on the range of the calibration data. r or R^2 near 1.00 (>0.990) indicate a linear relationship, but lower values do not distinguish between nonlinearity and variability [15].

Specificity. The application of qualitative RS for identification is achieved by comparing a sample spectrum to a reference spectrum or spectra either visually or using of a chemometric model. The specificity of the identification procedure is confirmed by using positive and negative controls. Materials to demonstrate specificity should be selected based on sound scientific judgment and risk specific to the application (e.g. materials handled in the same area). For quantitative applications, confirming of specificity against a physical or chemical property of interest is based on interpreting both Raman spectral attributes and chemometric parameters in terms of the intended application. The demonstration of specificity may include the following: spectral regions, wavelengths used by regression analysis for the calibration, variation in spectra from samples. The variation in material composition/sample matrix should be shown to have no significant effect on quantification of the property of interest within the specified procedure range as part of robustness studies [15].

Robustness. For robustness, changes in sample position, laser–sample interaction and hardware settings should be evaluated [15].

6.3 Chemometrics

MS can result in a high number of absorbance values even for single measurements, therefore specific statistical tools called chemometrics are used. The International Chemometrics Society (ICS) has defined chemometrics as 'the science of relating chemical measurements made on a chemical system to the property of interest (such as concentration) through the application of mathematical or statistical methods' [30]. Chemometrics is widely applied to the chemical data of molecular spectra, in particular if dealing with big data. Some reviews on chemometrics for specific fields exist. Some chemometrics techniques are typically applied and can provide an alternative way to analyze complex spectra data, including pre-processing spectral treatment, exploratory data analysis, discrimination of samples, classification analysis and multivariate calibrations [35]. Spectral processing treatments are typically used to give better performance modeling, which includes mean centering, spectral derivatization techniques and multiplicative correction, which are normally used in molecular spectra before being subject to chemometrics analysis [36].

Different chemometrics techniques applied for of classification purposes, such as classification between caries and non-caries teeth, are widely applied in dentistry. The chemometrics techniques applied for classification are exploratory data analyses, such as principal component analysis (PCA), unsupervised pattern recognition and supervised pattern recognition techniques, as shown in figure 6.1. Exploratory data analysis and unsupervised pattern recognition are commonly used to simplify groups of samples by reducing the amount of original data and gaining better knowledge of chemical data sets. Therefore, the main challenge of these is to remove the redundancy and noise while retaining the meaningful information contained in the original data [37].

Exploration data analysis (EDA) is a data reduction technique which defines a number of latent variables used to make linear combinations of the original variables. PCA is considered to be EDA. The pattern recognition techniques either unsupervised or supervised are among the widely used branches of chemometrics techniques. Supervised pattern recognition (or classification) relies on a grouping variable (class membership information) to estimate and assign class memberships, while unsupervised techniques work without this information to find the dominant patterns and outliers in the model or dataset [38].

Unsupervised pattern recognition can be differentiated from EDA because the aim of pattern recognition is to detect similarities, whereas EDA has no particular preconception as to whether or how many groups will be found. The unsupervised pattern recognition techniques include clustering analysis (CA) and similarity analysis (SA) [39]. Supervised pattern recognition (SPR) tries to make the class membership of the evaluated objects to a certain group known as training sets. It enables classifying new unknown samples (test samples) in one of the known classes on the basis of the variables used. SPR can be differentiated into class-modeling

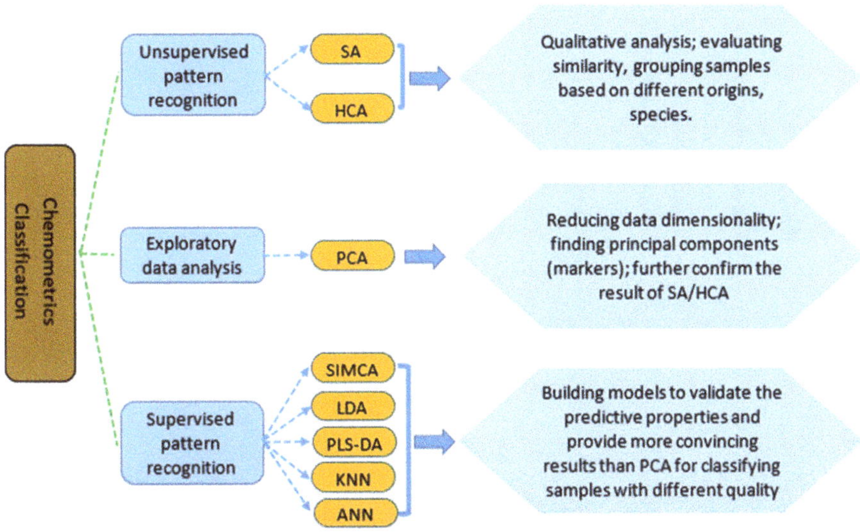

Figure 6.1. The chemometrics classification techniques widely applied for the identification of objects. SA = similarity analysis, HCA = hierarchical clustering analysis, PCA = principal component analysis, SIMCA = soft independent modeling of class analogy, LDA = linear discriminant analysis, PLS-DA = partial least-squares discriminant analysis, KNN = k-nearest neighbors, and ANN = artificial neural networks. (Reproduced from [2] under the terms and conditions of the Creative Commons Attribution (CC BY 4.0) license (http://creativecommons.org/licenses/by/4.0/).)

methods and discrimination methods. The most-used class-modeling method is known as soft independent modeling of class analogy (SIMCA). SIMCA considers the objects (samples) that fit the class model for a category as part of the class model and classify as non-members those that do not [40]. Discrimination models include linear discriminant analysis (LDA), partial least-squares discriminant analysis (PLS-DA), artificial neural networks (ANNs) and *k*-nearest neighbors (KNNs). PLS-DA is intended to find the variables and directions in the multivariate space capable of discriminating the established classes in the calibration set. ANNs consider finding the most appropriate grouping of training, learning and transfer function for classifying the data sets (variables) with a growing number of features and classified sets. In a simple form, ANNs try to imitate the operation of neurons in the brain. KNN is one of the most popular classification techniques based on distance algorithms. KNN is based on measuring the distances between the training samples and test samples to determine the final classification output [41].

6.4 Application of MS for the analysis of dental materials and for the quality of dental preparation

Summaries of recent publications (2015–2021) on the application of UV–vis and other molecular spectroscopic techniques, namely FS, MIRS, NIRS and RS, in dentistry are presented in table 6.1. Generally, these spectroscopic methods can be applied for direct diagnostic purposes, the quantification of certain compounds, the

Table 6.1. The use of MS in combination with statistical and chemometrics techniques in dentistry.

MS	Wavelength/ wavenumbers	Analytes/samples	Issues	Sample preparation	Statistical and chemometric techniques	Results	Ref.
FS	Excitation: 335–425 nm, interval of 10 nm; emission: 400–700 nm, cutoff 430 nm	Seven strains of oral cavity's bacteria	Determination of the fluorescence characteristics of the seven bacteria	A suspension of harvested bacteria in water was pipetted into a fluorescence cuvette.	MVA/PCA	The seven bacteria can be identified and classified using a combination of FS and PCA.	[43]
Fluorescence microscopy combined with FS and RS	FS—excitation: 380–450; emission: 450–800 nm. RS—excitation (laser): 532 nm	Extracted teeth for caries, periodontics, or orthodontics	Revealing the presence of protoporphyrin IX in carious enamel and dentin using FS and micro-RS	Teeth were selected and cross-sectioned using an Isomet diamond saw. Each tooth cross-section was kept in a separate box in ultrapure water at 4 °C.	NA	By comparing the spectra (FS and RS), protoporphyrin IX can be detected in enamel and dentin.	[44]
FS	Excitation: 365 nm; emission: 400–600 nm	Fifty extracted, non-damaged and intact bovine incisors	Evaluating the effect of whitening using H_2O_2 on the fluorescence and color of the enamel	A single enamel or dentin specimen, 6 mm in diameter and 1 mm high, was obtained from the labial surface of	ANOVA, Tukey's test, t-test	Whitening by H_2O_2 induced a significant decrease in the fluorescence intensity of tooth dentin, and	[45]

(*Continued*)

Table 6.1. (*Continued*)

MS	Wavelength/ wavenumbers	Analytes/samples	Issues	Sample preparation	Statistical and chemometric techniques	Results	Ref.
			and dentin substrates	each crown using a trephine mill.		promoted significant color changes in dentin and enamel.	
UV–vis	NA	Self-prepared nicotine stained dental enamel fragments ($n = 36$)	Evaluating potential ingredients of tooth bleaching agents on the color stability of nicotine staining in the dental enamel model	Samples were subjected to the bleaching ingredients (30 min); analysis was performed to evaluate the color stability using a colorimeter and UV–vis spectrometer.	ANOVA	Limonene, coconut diethanolamide and carbamide peroxide seemed to be more discoloration active for the nicotine stained dental enamel model.	[46]
UV–vis, FS	Excitation: 350–700 nm; emission: 450–750 nm	Commercially composite resins; dental tissue specimens	Evaluating the fluorescence intensity of different composite resins and comparing those values with the fluorescence intensity of dental tissues	The composite resin specimens were excited using an UV LED.	Dunnet, Tukey and Kruskall–Wallis tests	Significant differences in fluorescence among the analyzed composite resins and the dental tissues were observed.	[21]
FS	Excitation (laser): 632.8 and	Enamel model containing nAlPc (10 mg/L), protelan	Studying the interaction of laser radiation using	Fluorescence measurement (using fiber probe	NA	nAlPc can be applied for detecting enamel	[47]

	335 nm; emission: 625–750 nm	(1%) and some additional components	aluminum phthalocyanine nanoparticles (nAlPc) on the enamel surface for photothrombotics of initial caries	optic) in samples containing nAlPc and surfactant was carried out at the following time intervals: 0–180 minutes, 1–6 days.		microcracks and pathogenic microflora accumulations.	
IR, UV–vis	4000–600 cm^{-1}; 200–2000 nm	Self-prepared collagen using some active compound and cross-linking agent	Predicting the characterization of the collagens using IR and UV–vis	IR spectra were recorded using ATR; UV–vis were recorded in gel form.	NA	IR and UV–vis spectra can predict the presence of the active compounds and cross-linking agent.	[48]
Nano UV–vis	224 nm	Randomly chosen implants from eight different processes batches	Investigating the release of a doxycycline coating layer on titanium-zirconium alloy surfaces	Released studies of doxycycline were performed in some buffer solutions. All samples were transferred to Eppendorf tubes and 0.5 ml release buffer (pH 3) was added; tests at 300 rpm at room temperature (1 h, 3 h, 6 h, 24 h, 72 h, and 7 days).	*t*-test ($p < 0.05$) (the calibration curve was developed 0.005–0.25 mg ml^{-1}; $R^2 = 0.993/0.8673$; other validation parameters were not reported)	A significant release of doxycycline was observed after 24 h incubation.	[49]
FS	Excitation: 408 nm		Studying the effects of demineralization		ANOVA, PCA	Demineralization leading to loss of	[50]

(Continued)

Table 6.1. (*Continued*)

MS	Wavelength/ wavenumbers	Analytes/samples	Issues	Sample preparation	Statistical and chemometric techniques	Results	Ref.
	and 532 nm; emission: 400–1000 nm	12 lower incisor teeth of adult male Wistar rats	on teeth and bones evaluated by FS, and micro energy-dispersive X-ray fluorescence spectrometry	The probe was placed on the teeth perpendicularly.		tissue quality may be assessed by FS in combination with PCA analysis.	
UV–vis	220 nm	Methyl methacrylate (MMA)	Analysis of residual MMA from the elution of different specimen fabrication of polymethyl methacrylate (PMMA)	The specimens were ground or polished, then ultrasonicated in distilled water, then incubated for 12 h. The two storing conditions were water (37 °C) and air (20 °C).	Kolmogorov–Smirnov test (the range of the calibration curve was reported, R^2 was missing; other parameters were not reported)	Fabrication procedures (pouring or injection), polishing, pressure during polymerization, and storage medium affected the MMA elution.	[51]
FS	Excitation (laser): 337.1 nm; emission: 363–627 nm	Three samples of dental caries and one sample of sound teeth	Distinguishing the emission spectra of dental caries and sound tooth	Each tooth from the samples was analyzed by using a sample holder.	NA	Different spectra were observed visually between the sound tooth and those of different carious stages.	[52]
Micro-RS and FS	RS: 400–1400 cm^{-1}. FS—excitation	Different parts of the tooth including healthy and dentin	Diagnostics of pathologies of the oral cavity organs	No detailed methods were reported.	NA	Different RS and FS spectra were observed visually between sound	[53]

	(laser): 514.5 nm; emission: 550–800 nm	caries damaged areas	using non-invasive optical methods			dentin, first point of infected dentin and second point of infected dentin.	
UV–vis, CD, ATR-FT-IR	UV: 200–340 nm. CD: 190–250 nm. IR: 1700–1600 cm^{-1}	Extracted protein from healthy human teeth samples	Characterization of protein extracted from four types of human teeth (molar, premolar, canine, and incisor)	Washed teeth were crushed in liquid N_2, powdered teeth were mixed with demineralization solution and protease inhibitors (20 °C, 2 d), dialyzed (HEPES, pH 7.4), then supernatant was concentrated using membrane base cut filter (3.5 kDa).	Student's unpaired t-test	This work showed that most protein existed in an unstructured state and the structural homogeneity of proteins in different teeth was observed.	[54]
FS, ATR-FT-IR	FS—excitation (laser): 405 nm, emission: 450–800 nm. FT-IR: 7800–350 cm^{-1}	Teeth of stage II, III and IV based on a visual and tactile inspection ($n = 20$)	Applying auto-fluorescence and FT-IR spectra to diagnose and stage caries	FS: The laser was focused on the center of each lesion for one second using a convex lens ($f = 10$ cm). The spot size on the lesion was less than 10 μm. FT-IR: 2 mg of	ANOVA followed by a Tukey's post-hoc test	FS: The spectral emission slope at 550–600 nm showed the best match with the caries specimens. FT-IR: The intensities of the bands 1550, 1460–1415, 870,	[55]

(Continued)

Table 6.1. (*Continued*)

MS	Wavelength/ wavenumbers	Analytes/samples	Issues	Sample preparation	Statistical and chemometric techniques	Results	Ref.
				ground powder of a carious lesion was mixed with 10 mg KBr and made into a thin film for the measurements.		1090–1032, 960, 2930, 2860 and 3700–3400 cm^{-1} were decreased gradually as caries progressed.	
UV–vis	225 nm	Chlorhexidine gluconate (CHX)	Studying the ability of two types of polyelectrolyte multilayer thin film to control the loading and release of CHX and their toxicity to anaerobic bacteria	Loaded thin films with CHX were dipped into H_2O_2 solution to release all CHX from the quartz slide.	NA	Calibration curve: 0.0005–0.0025 M, $R^2 = 0.9931$, other validation parameters were not reported. The results indicated that the two types of thin films released 0.00052 and 0.00071 M CHX, respectively.	[56]
UV–vis, micro-RS	UV–vis: 200–1100 nm. RS: 300–3300 cm^{-1}	Four different commercial Giomer materials in the field of restorative dentistry	Investigation of the morphological and structural characteristics of Giomers in order to make a comparative analysis	Polyurethane stamps with four negative disk specimen shapes were prepared and filled with test materials. A transparent matrix strip was	NA	Giomers show very similar behavior to the other restorative materials investigated in this study.	[57]

				placed on the surface restorative materials. RS: fixing the samples on a standard glass slide; 532 nm, exposure time 1 s (30 accumulations). UV–vis: no detailed method was reported.			
FS	Excitation (laser): 445 nm; emission: 450–780 nm	*In vivo*: 60 patients with abfraction in the cervical area and 60 patients with early erosion.	Using spectroscopic methods to study the dental hard tissue erosion and abfraction	Fluorescence spectra were recorded using of a patented device designed based on a fiber-optic spectrometer. The probing area of the teeth was determined by the area of the waveguide and was equal to 0.28 mm^2.	NA	Laser induced FS showed its effectiveness in the diagnosis of abfraction and erosion, but has some limitations. NN recognition algorithms and the support vector machine method were needed to help the diagnostics.	[58]
Mid-IR	The combined region of 4500–580, 6200–8000	Upper-premolar teeth	Qualitative and quantitative analysis of dental caries	The sample specimens were subjected to a diffuse reflectance NIR spectrometer in a wavenumber	PCA for qualitative and multivariate calibration (PCR and PLS) for	PCA using PC1 and PC2 accounting for 98% could classify caries and non-caries teeth. PLS	[59]

(*Continued*)

Table 6.1. (*Continued*)

MS	Wavelength/ wavenumbers	Analytes/samples	Issues	Sample preparation	Statistical and chemometric techniques	Results	Ref.
	and 10 000 −10 500 cm^{-1}			range of 4000–12 500 cm^{-1}.	quantitative analysis	using the first derivative pretreatment spectra could predict enamel surface changes and the crystallite size most accurately and precisely.	
Mid-IR	Mid-infrared region of 4000–400 cm^{-1}	Dental samples (powder, paste, gel and liquid)	Detection of gelatin in dental materials using ATR-FT-IR	Dental paste containing bovine and porcine gelatins with different concentrations was scanned using attenuated total reflectance (ATR) within wavenumbers of 4000–400 cm^{-1} with a resolution of 2 cm^{-1} and 32 scans.	NA	The gelatins were present in 9 of 42 samples of dental materials but the species of origin was not confirmed. ATR-FT-IR can be applied as a reliable tool to detect gelatin in dental materials and other pharmaceuticals.	[60]

Mid-IR	Mid-infrared region of 4000–400 cm^{-1}	Dental materials	Identification of gelatin sources in dental materials	The gelatin samples in dental materials were recorded using ATR at 4000–500 cm^{-1} with a resolution of 4 cm^{-1} and 32 scans.	SIMCA and PCA for classification and differentiation	FT-IR spectra at wavenumbers of 570–1662 cm^{-1} combined with PCA and SIMCA are successfully applied for identification of gelatin sources in dental material products. Four of the 49 dental materials used in dental hospitals contain gelatin and tend to be similar to pork gelatin, which is a non-halal material.	[61]
RS	Peak at 959 cm^{-1}	Six anterior healthy teeth, preserved in a 0.5% (w/w) chloramine solution	Analysis of dental enamel treated with an over-the-counter whitening product	Raman spectra of samples were obtained using a near-infrared laser (785 nm) at 400–1500 cm^{-1} with a spectral resolution of 4 cm^{-1}.	One-way ANOVA	Raman spectra using a peak corresponding to the PO_4^{3-} symmetric stretching band showed that a maximum of demineralization occurred after	[61]

(Continued)

Table 6.1. (*Continued*)

MS	Wavelength/ wavenumbers	Analytes/samples	Issues	Sample preparation	Statistical and chemometric techniques	Results	Ref.
						application of a whitening product between days 27 and 34.	
High-resolution RS	Peak at around 960 cm^{-1}	Human teeth from the Department of Dentistry at University Hospital, Kyoto	Analysis of subtle structural changes occurring in dental enamel and dental caries	The samples were measured using a Raman spectrometer using a green laser beam (532 nm) as the excitation source.	NA	RS could detect the dental caries in comparison to other methods.	[32]

FS: Fluorescence spectroscopy; RS: Raman spectroscopy; MIR: mid-infrared; CD: circular dichroism; ANOVA = analysis of variance; PCA = principal component analysis; SIMCA = soft independent modeling class analogy; NA: not available or not reported.

evaluation of some techniques, the determination of some oral bacteria, and direct detection or characterization of certain compounds on enamel, dentin or whole teeth. Some reviews have existed regarding the application of molecular spectroscopies in dentistry, such as reviews on the application of RS for the analysis of tooth structure [42].

To obtain reliable results for any chemical measurements (qualitative and quantitative), all the spectroscopic methods used should be first validated prior routine applications, and all validation parameters should meet the pre-determined acceptance criteria (see section 6.2). Spectroscopic methods can be applied in research work for dentistry for the purposes shown in table 6.1. Unfortunately, most of the publications in table 6.1 did not describe or report their methods of validation in detail. Without complete reported validation data, their conclusions cannot be verified in terms of their validity, and a as consequence they have only limited value for the reader. Many of the publications also did not report their statistical methods in detail. The authors recommend that researchers first perform validation methods before applying any published methods, as required by the Pharmacopeia. The detailed validation methods can follow the current Pharmacopeia of each country.

References

[1] Breuer D 2007 Molecular spectroscopy in the ultraviolet and visible range *The MAK-Collection for Occupational Health and Safety* (New York: Wiley) pp 1–15

[2] Rohman A and Windarsih A 2020 The application of molecular spectroscopy in combination with chemometrics for halal authentication analysis: a review *Int. J. Mol. Sci.* **21** 1–18

[3] Mortensen D, Dannemand K, Twetman S and Keller M K 2014 Detection of non-cavitated occlusal caries with impedance spectroscopy and laser fluorescence: an *in vitro* study *Open. Dent. J.* **8** 28–32

[4] Rohman A and Putri A R 2019 The chemometrics techniques in combination with instrumental analytical methods applied in halal authentication analysis *Indones. J. Chem.* **19** 262–72

[5] Wenzel T and Dana C A 2020 1.1: Introduction to molecular spectroscopy *Chemistry LibreTexts* https://chem.libretexts.org/Bookshelves/Analytical_Chemistry/Book%3A_Molecular_and_Atomic_Spectroscopy_(Wenzel)/1%3A_General_Background_on_Molecular_Spectroscopy/1.1%3A_Introduction_to_Molecular_Spectroscopy_ (Accessed: 10 March 2021)

[6] Smith B C 2011 *Fundamentals of Fourier Transform Infrared Spectroscopy* 2nd edn (Boca Raton, FL: CRC Press) pp 1–17

[7] The United States Pharmacopeial Convention 2020 <857> Ultraviolet–Visible Spectroscopy *USP 43–NF 38* (North Bethesda, MD: The United States Pharmacopeial Convention) pp 1–12

[8] The United States Pharmacopeial Convention 2020 <1857> Ultraviolet–Visible Spectroscopy, Theory and Practice *USP 43–NF 38* (North Bethesda, MD: The United States Pharmacopeial Convention) pp 1–7

[9] The United States Pharmacopeial Convention 2020 <853> Florescence Spectroscopy *USP 43–NF 38* (North Bethesda, MD: The United States Pharmacopeial Convention)

[10] The United States Pharmacopeial Convention 2020 <1853> Florescence Spectroscopy, Theory and Practice *USP 43–NF 38* (North Bethesda, MD: The United States Pharmacopeial Convention)

[11] The United States Pharmacopeial Convention 2020 <856> Near-Infrared Spectroscopy *USP 43–NF 38* (North Bethesda, MD: The United States Pharmacopeial Convention)

[12] The United States Pharmacopeial Convention 2020 <1856> Near-Infrared Spectroscopy, Theory and Practice *USP 43–NF 38* (North Bethesda, MD: The United States Pharmacopeial Convention)

[13] The United States Pharmacopeial Convention 2020 <854> Mid Infrared Spectroscopy *USP43-NF38* (North Bethesda, MD: The United States Pharmacopeial Convention)

[14] The United States Pharmacopeial Convention 2020 <1854> Mid Infrared, Theory and Practice *USP 43–NF 38* (North Bethesda, MD: The United States Pharmacopeial Convention)

[15] The United States Pharmacopeial Convention 2020 <858> Raman Spectroscopy *USP 43–NF 38* (North Bethesda, MD: The United States Pharmacopeial Convention)

[16] The United States Pharmacopeial Convention 2020 <1858> Raman Spectroscopy Theory and Practice *USP 43–NF 38* (North Bethesda, MD: The United States Pharmacopeial Convention)

[17] The United States Pharmacopeial Convention 2020 <1958> Analytical Instrument Qualification *USP 43–NF 38* (North Bethesda, MD: The United States Pharmacopeial Convention)

[18] The United States Pharmacopeial Convention 2020 <1225> Validation of Compendial Procedures *USP 43–NF 38* (North Bethesda, MD: The United States Pharmacopeial Convention)

[19] Indrayanto G 2018 Validation of chromatographic methods of analysis: application for drugs that derived from herbs *Profiles of Drug Substances, Excipients and Related Methodology* (New York: Academic) pp359–92

[20] The United States Pharmacopeial Convention 2020 <1853> Florescence Spectroscopy, Theory and Practice *USP 43–NF 38* (North Bethesda, MD: The United States Pharmacopeial Convention)

[21] Da Silva T M, De Oliveira H P M, Severino D, Balducci I, Huhtala M F R L and Gonçalves S E P 2014 Direct spectrometry: a new alternative for measuring the fluorescence of composite resins and dental tissues *Oper. Dent.* **39** 407–15

[22] Charvát J, Procházka A, Fričl M, Vyšata O and Himmlová L 2020 Diffuse reflectance spectroscopy in dental caries detection and classification *Signal Image. Video. Process.* **14** 1063–70

[23] The United States Pharmacopeial Convention 2020 <1039> Chemometrics *USP 43–NF 38* (North Bethesda, MD: The United States Pharmacopeial Convention)

[24] Kim I H, Son J S, Min B K, Kim Y K, Kim K H and Kwon T Y 2016 A simple, sensitive and non-destructive technique for characterizing bovine dental enamel erosion: attenuated total reflection Fourier transform infrared spectroscopy *Int. J. Oral. Sci.* **8** 54–60

[25] The United States Pharmacopeial Convention 2020 <197> Spectroscopy Identification Test *USP 43–NF 38* (North Bethesda, MD: The United States Pharmacopeial Convention)

[26] Indrayanto G and Rohman A 2020 The use of FTIR spectroscopy combined with multivariate analysis in food composition analysis *Spectroscopic Techniques and Artificial Intelligence for Food and Beverage Analysis* (Singapore: Springer) pp 25–51

[27] Aminoroaya A, Esmaeely Neisiany R, Nouri Khorasani S, Panahi P, Das O and Ramakrishna S 2020 A review of dental composites: methods of characterizations *ACS Biomater. Sci. Eng.* **6** 3713–44

[28] Hedzelek W, Wachowiak R, Marcinkowska A and Domka L 2008 Infrared spectroscopic identification of chosen dental materials and natural teeth *Acta Phys. Pol.* A **114** 471–84
[29] Wulfman C, Sadoun M and Lamy De La Chapelle M 2010 Interest of Raman spectroscopy for the study of dental material: the zirconia material example *IRBM* **31** 257–62
[30] Timchenko E, Timchenko P, Volova L, Frolov O, Zibin M and Bazhutova I 2020 Raman spectroscopy of changes in the tissues of teeth with periodontitis *Diagnostics* **10** 1–12
[31] Marin E *et al* 2020 Raman spectroscopy for early detection and monitoring of dentin demineralization *Dent. Mater.* **36** 1635–44
[32] Miyamoto N *et al* 2020 Molecular fingerprint imaging to identify dental caries using Raman spectroscopy *Materials* **13** 1–15
[33] Gordon L M, Cohen M J, MacRenaris K W, Pasteris J D, Seda T and Joester D 2015 Amorphous intergranular phases control the properties of rodent tooth enamel *Science* **347** 746–50
[34] Buchwald T and Buchwald Z 2019 Assessment of the Raman spectroscopy effectiveness in determining the early changes in human enamel caused by artificial caries *Analyst* **144** 1409–19
[35] Singh I, Juneja P, Kaur B and Kumar P 2013 Pharmaceutical applications of chemometric techniques *ISRN Anal. Chem.* **2013** 1–13
[36] Nunes C A 2014 Vibrational spectroscopy and chemometrics to assess authenticity, adulteration and intrinsic quality parameters of edible oils and fats *Food Res. Int.* **60** 255–61
[37] Berrueta L A, Alonso-Salces R M and Héberger K 2007 Supervised pattern recognition in food analysis *J. Chromatogr.* A **1158** 196–214
[38] Miller J N and Miller J C 2010 *Statistics and Chemometrics for Analytical Chemistry* 6th edn (Harlow: Prentice Hall) p 278
[39] Rácz A, Gere A, Bajusz D and Héberger K 2010 Is soft independent modeling of class analogies a reasonable choice for supervised pattern recognition? *RSC Adv.* **8** 10–21
[40] Ali N, Neagu D and Trundle P 2019 Evaluation of *k*-nearest neighbour classifier performance for heterogeneous data sets *SN Appl. Sci.* **1** 1–15
[41] Ramakrishnaiah R *et al* 2015 Applications of Raman spectroscopy in dentistry: analysis of tooth structure *Appl. Spectrosc. Rev.* **50** 332–50
[42] Kang S, Josselin de Jong E, Higham S M, Hope C K and Kim B 2020 Fluorescence fingerprints of oral bacteria *J. Biophotonics* **13** e201900190
[43] Slimani A *et al* 2016 Porphyrin and pentosidine involvement in the red fluorescence of enamel and dentin caries *Int. J. Exp. Dent. Sci.* **5** 1–10
[44] Caneppele T M F, Torres C R G and Bresciani E 2015 Analysis of the color and fluorescence alterations of enamel and dentin treated with hydrogen peroxide *Braz. Dent. J.* **26** 514–8
[45] Lertsukprasert N and Locharoenrat K 2020 Efficiency of tooth bleaching agent on staining and discoloration characteristics of nicotine stained dental enamel model *BMC Oral. Health* **20** 221
[46] Zolotareva J O, Farrakhova D S, Kupriyanova E N and Loschenov V B 2018 Aluminum phthalocyanine nanoparticles activation for local fluorescence spectroscopy in dentistry *Biomed. Photonics* **7** 4–20
[47] Zgârian R G, Tihan G T and Barbaresso R C 2016 Rəu I. Spectral characterization of some collagen based composite for dental application *UPB Sci. Bull.* B **78** 99–110
[48] Rahmati M *et al* 2020 Coating doxycycline on titanium-based implants: two *in vivo* studies *Bioact. Mater.* **5** 787–97

[49] Paolillo F R *et al* 2016 Fluorescence spectroscopy of teeth and bones of rats to assess demineralization: *in vitro*, *in vivo* and *ex vivo* studies *J. Photochem. Photobiol.* B **165** 291–7

[50] Keul C, Seidl J, Güth J F and Liebermann A 2020 Impact of fabrication procedures on residual monomer elution of conventional polymethyl methacrylate (PMMA)—a measurement approach by UV/vis spectrophotometry *Clin. Oral. Investig.* **24** 4519–30

[51] Abdel-Rahman Saeed Marouf A and Awad Khairallah Y 2019 Photoemission spectra of sound tooth and those of different carious stages *Eur. J. Biophys.* **7** 23

[52] Goloshchapov D L, Kashkarov V M, Ippolitov Y A, Prutskij T and Seredin P V 2018 Early screening of dentin caries using the methods of micro-Raman and laser-induced fluorescence spectroscopy *Results Phys.* **10** 346–7

[53] Sharma V *et al* 2019 Characterization of protein extracts from different types of human teeth and insight in biomineralization *Sci. Rep.* **9** 1–9

[54] Ko C C, Yi D H, Lee D J, Kwon J, Garcia-Godoy F and Kwon Y H 2017 Diagnosis and staging of caries using spectral factors derived from the blue laser-induced autofluorescence spectrum *J. Dent.* **67** 77–83

[55] Wacharanad S, Sasimomthon W, Wongyai P, Vudhivanich A and Tippawan K 2016 Activity of chlorhexidine gluconate loaded at varying polyelectrolyte multilayers against aggregatibacter actinomycetemcomitans *MATEC Web Conf.* **77** 11003

[56] Condò R *et al* 2017 A deep morphological characterization and comparison of different dental restorative materials *Biomed. Res. Int.* **2017** 7346317

[57] Sarycheva I, Yanushevich O and Minakov D 2020 Diagnostics of non-carious lesions of dental hard tissues with the methods of optical spectroscopy and radiography *Braz. Dent. Sci.* **23** 8

[58] Sodata P and Peerapattana J 2020 Application of near infrared spectroscopy with chemometrics for qualitative and quantitative dental caries assessment *Vib. Spectrosc.* **111** 103170

[59] Irfanita N, Jaswir I, Mirghani M E S, Sukmasari S, Ardini Y D and Lestari W 2017 Rapid detection of gelatin in dental materials using attenuated total reflection Fourier transform infrared spectroscopy (ATR-FTIR) *J. Phys.: Conf. Ser.* **884** 12090

[60] Roswiem A P and Mustaqimah D N 2020 Identification of gelatin source in toothpaste products using combination of attenuated total reflection-Fourier transform infrared (ATR-FTIR) spectroscopy and chemometrics *Int. J. Halal. Res.* **2** 30–9

[61] Silveira J *et al* 2018 Raman spectroscopy analysis of dental enamel treated with whitening product—influence of saliva in the remineralization *Spectrochim. Acta* A **198** 145–9

www.ingramcontent.com/pod-product-compliance
Ingram Content Group UK Ltd.
Pitfield, Milton Keynes, MK11 3LW, UK
UKHW051849100926
8556IPUK00005B/530